W A Gruver & E Sachs

North Carolina State University/Technische Universität Berlin

# Algorithmic methods in optimal control



π

Pitman Advanced Publishing Program

BOSTON · LONDON · MELBOURNE

PITMAN PUBLISHING LIMITED
39 Parker Street, London WC2B 5PB

PITMAN PUBLISHING INC
1020 Plain Street, Marshfield, Massachusetts

*Associated Companies*
Pitman Publishing Pty Ltd., Melbourne
Pitman Publishing New Zealand Ltd., Wellington
Copp Clark Pitman, Toronto



AMS Subject Classifications: 49DXX, 65KXX, 93CXX, 90CXX

Library of Congress Cataloging in Publication Data
Gruver, W A
Algorithmic methods in optimal control.
(Research notes in mathematics; 47)
Bibliography: p.
1. Control theory. 2. Mathematical optimization.
3. Algorithms. I Sachs, E., joint author.
II. Title. III. Series.
QA402.3.G77 629.8'312 80-21442
ISBN 0-273-08473-9



ISBN 0 273 08473 9

Printed in Great Britain by
Biddles Ltd, Guildford, Surrey

# Preface

Our purpose in writing this book is to provide the reader with a detailed treatment of selected mathematical theory and applications of computational methods for optimal control. We shall consider techniques that can be viewed as extensions of nonlinear programming methods for the control of dynamic systems described by ordinary and partial differential equations. The reader should be prepared in functional analysis and differential equations. Otherwise, the treatment is self-contained; knowledge of control theory and optimization is not assumed. The contents of this book are suitable for self-study or for use in a graduate course in applied mathematics, and the allied fields of engineering, operations research, statistics, and computer science.

This book is not intended to be an exaustive treatment of either optimal control theory or optimization methods. Instead, we emphasize extensions of a family of quasi-Newton methods that have been demonstrated to be very effective in practice. Our treatment of constrained optimal control is confined to the use of projection in Hilbert space. The interested reader is encouraged to consult the references cited at the end of each chapter for additional details and the vast literature on optimal control, nonlinear programming, and approximation theory.

The first chapter contains introductory remarks and some motivational examples. In Chapter 2 we discuss inexact step length rules and their relationship to the Zoutendijk condition. The subsequent three chapters are devoted to quasi-Newton methods, including the construction of approximations to the inverse Hessian and results on global convergence and convergence rates. In Chapter 6 we discuss conjugate gradient methods as derived from quasi-Newton methods. Chapters 7 and 8 are devoted to conditional gradient and projection methods for constrained optimal control. Chapter 9 deals with approximation-type methods that are also applicable to optimal control problems. Chapters 10, 11 and 12 present results in the control of linear and non-linear systems described by partial differential equations, including the use of algorithmic methods to compute optimal controls. The last four chapters are a selection of applications of optimal control to orbital rendezvous, stochastic service systems, production planning, and biomechanical models for the evaluation of manual lifting tasks. These application studies provide detailed numerical results that were obtained using quasi-Newton and gradient projection algorithms. The Habilitation thesis of one of the authors (E. W. S.) has been integrated into those parts of the book dealing with quasi-Newton methods and the application of optimal control to nonlinear systems described by partial differential equations.

The technical advice and encouragement of Professor W. Krabs has been a vital stimulus to each of us during the preparation of this work. Support from the Alexander von Humboldt Stiftung, the Deutsche Akademische Austauschdienst, the Deutsche Forschungsgemeinschaft, the DFVLR Institut für Dynamik der Flugsysteme, and the National Science Foundation is gratefully acknowledged by the authors.

Finally, we thank W. Eichenauer and Dr. J. Jahn for improvements to the manuscript; Dr. M. A. Ayoub, N. H. Engersbach, Dr. C. F. Klein, M. B. Muth, and Dr. S. L. Narasimhan for their contributions to Chapters 13- 16; and Mrs. H. Schmidt for her excellent typing of the final manuscript.

Torrance, California W. A. G.
Berlin, West Germany E. W. S.
August 1980

# Contents

# 1 Introduction

## 1.1 DEFINITION OF THE PROBLEM

Let B be a Banach space, U a nonvoid subset of this space and F a functional on U. Then we can define the optimization or minimization problem: minimize F(u) on U. F is called the objective or cost functional and U the constraint or feasible set. We call an element $\hat{u} \in U$, a solution, minimal point or optimal point of the optimization problem if and only if

$$F(\hat{u}) \leq F(u) \qquad \text{for all } u \in U \ . \tag{1.1}$$

The number $\inf \{F(u) : u \in U\}$ is called the minimal or optimal value. Most of the problems in optimal control or best approximation can be expressed in the form (1.1), an infinite dimensional optimization problem.

If U is a nontrivial subset of B, then (1.1) is called a constrained optimization problem; and if U = B, then (1.1) is an unconstrained optimization problem.

Let B, C be two normed spaces and G a mapping

$$G : B \to C \ .$$

Then G is called Fréchet differentiable at $\hat{u} \in B$ if F is defined on a neighborhood of $\hat{u}$ and if there exists a continuous linear mapping $F'_{\hat{u}} : B \to C$ such that

$$\lim_{||v|| \to 0} ||F(\hat{u} + v) - F(\hat{u}) - F'_{\hat{u}}(v)|| \ \ ||v||^{-1} = 0 \ .$$

We say that F is Fréchet differentiable on a set if it is Fréchet differentiable at each point of this set. If F is a functional on a Hilbert space H, then $F'_{\hat{u}}$ is a linear continuous functional on H, i.e. an element of the dual space $H^*$ and can be identified by an element of the underlying space H itself. This element is called the gradient $\nabla F(\hat{u})$ at the point $\hat{u}$.

A set U is called convex, if

$$u, v \in U, \quad \alpha \in [0,1] \Rightarrow \alpha u + (1-\alpha) v \in U$$

holds. Assume U is a convex set, then $F : U \to \mathbb{R}$ is called a convex functional, if

$$u, v \in U, \quad \alpha \in [0,1] \Rightarrow$$

$$F(\alpha u + (1-\alpha)v) \leq \alpha F(u) + (1-\alpha) F(v)$$

is true. If F is a continuous convex functional on U, then we can replace the derivative by a set of linear functionals. The subgradient $\partial F(\hat{u})$ of F at $\hat{u} \in U$ consists of all continuous linear functionals $\ell$ on B, such that for all $v \in U$

$$\ell(v - \hat{u}) \leq F(v) - F(\hat{u}) .$$

If F is Fréchet differentiable at $\hat{u}$, then $\partial F(\hat{u})$ consists only of $F'_{\hat{u}}$.

The basic necessary and sufficient condition for optimality is:

THEOREM 1.1: If $\hat{u} \in U$ is a solution of the constrained minimization problem and F is Fréchet differentiable at $\hat{u}$, then

$$F'_{\hat{u}}(u - \hat{u}) \geq 0 \qquad \text{for all } u \in U . \tag{1.2}$$

If F is convex on U, then (1.2) is also sufficient for the optimality of F. Suppose F is only continuous and convex, then

$$\Theta \in \partial F(\hat{u}) \tag{1.3}$$

is characteristic for optimality of $\hat{u}$.

## 1.2 ITERATIVE METHODS

For the actual solution of an optimization problem in many cases, a sequence of points is constructed such that the value of the objective is decreased and approaches a minimal value.

The basic idea is to start with some point $u_i$ and find a direction $v_i$ such that the value of the objective decreases in that direction, i.e.,

$$F'_{u_i}(v_i) < 0 \ , \tag{1.4}$$

and to proceed a certain step length $\alpha_i$ in that direction

$$u_{i+1} = u_i + \alpha_i v_i \ .$$

The selection of the step length can be important. If it is selected too large, then the function value at $u_{i+1}$ may not be smaller than at $u_i$. If it is too small, then the decrease may not be sufficient enough to come close to the minimal value during the iterations. Various rules are discussed in Chapter 2.

The determination of the direction of descent $v_i$ is a more delicate problem for which numerous approaches have been developed.

Let us suppose the problem is posed in a Hilbert space H. Then, at first glance, it is most intuitive to select $v_i$ such that the value in (1.4) becomes as small as possible. Since scalar multiplication of $v_i$ satisfies this requirement arbirarily, we select $v_i$ from a bounded set, the unit ball. Then

$$v_i = - \frac{\nabla F(u_i)}{\| \nabla F(u_i) \|} \ ,$$

the steepest descent direction. However, steepest descent is not always the best choice. For example, if the graph of the objective is "banana-shaped", then a steepest descent method may not even converge.

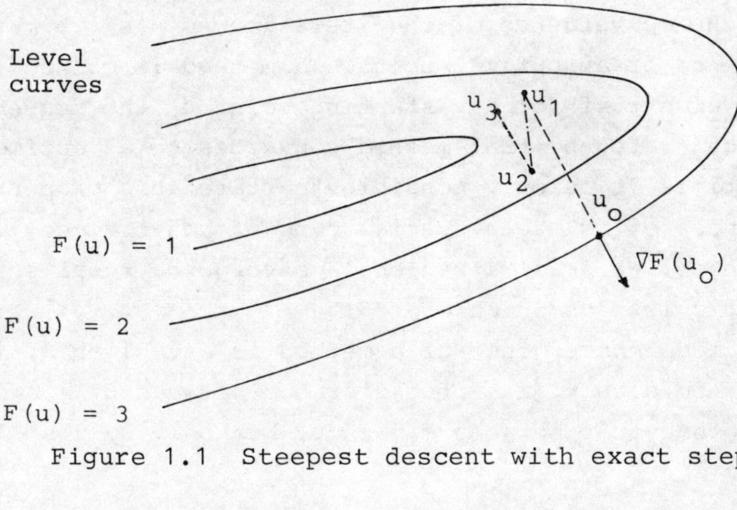


Figure 1.1 Steepest descent with exact step size

Since it is possible to measure the descent in the Euclidean inner product, use of another inner product may result in a different descent direction. With different inner products, circles become ellipsoids, etc., and the descent direction could point into the minimum.

Defining

$$\langle v,u\rangle_A = \langle v,Au\rangle$$

as an inner product, the "steepest" descent in this metric is

$$v_i = - \frac{A\nabla F(u_i)}{\| A\nabla F(u_i) \|} \ .$$

Methods of this type are called <u>variable metric methods</u>. Thus, one problem is a clever determination of A. In Chapter 3 we give a derivation of a class of updates based on the concept underlying Newton's method. These updates are called <u>quasi-Newton methods</u>.

The most important property of iterative methods is its convergence, i.e. does the sequence of iterative vectors $\{u_i\}_{\mathbb{N}}$ yield information on the optimal value, point, etc.

We say that $\{F(u_i)\}$ <u>converges globally</u> to the minimal value if

$$\lim_{i \to \infty} F(u_i) = \inf \{F(u) : u \in U\} \ . \tag{1.5}$$

In (1.5) we do not assume that a minimal point exists. To investigate the convergence of the iteration points, we must assume existence of an optimal point. A method is called <u>globally convergent</u> if for any starting point $u_o$ the sequence of iterates generated by that method converges to an optimal point. Sometimes it is only possible to state this property for subsequences or to use a special type of convergence, e.g. weak convergence. We deal with global convergence results for variable metric methods in Chapter 3.

Even if global convergence of a method is established, it may converge too slowly, i.e. the distance between $u_{i+1}$ and $\hat{u}$ is smaller than that between $u_i$ and $\hat{u}$, but only by a small amount. The "velocity" of the convergence is the convergence rate for which various definitions are used. The main idea is to estimate $\| u_{i+1} - \hat{u} \|$ by $\| u_i - \hat{u} \|$. Often this estimate can be obtained only if the starting point is sufficiently close to an optimal point. If this requirement must be met, we talk of local convergence results. For variable metric methods these questions are treated in Chapter 5.

There are various ways to update the matrix A at each iteration. In most cases, however, information about A usually must be stored for the next iteration. One popular approach, treated in Chapter 6, is known as conjugate gradient methods. They have the feature that for a quadratic cost functional the descent directions are orthogonal with respect to a certain inner product.

The previous paragraphs have treated unconstrained problems. If the optimization problem is constrained, in addition to the requirement that the values of the objective at the iterates decrease, we also have to ensure that the iterates lie in the feasible set.

By choosing the steepest descent direction in the feasible set, we obtain the Frank-Wolfe or conditional gradient method.

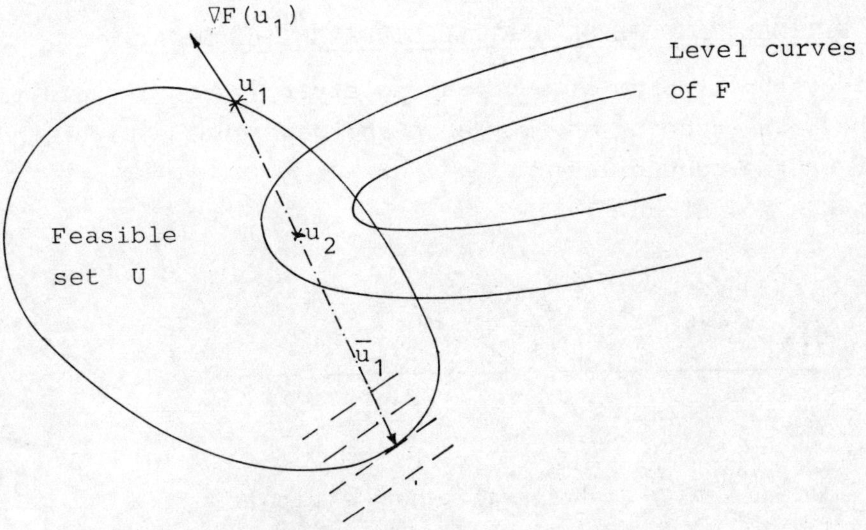


Figure 1.2 Conditional gradient method

In Chapter 7 we present some results on convergence and their rates for several step length rules. The application to control problems with control constraints is rather straightforward.

Chapter 8 treats two projection methods. The first incorporates a projection into the feasible set during the step size determination. The other projects a descent direction into the feasible set and determines a step size along the line between the projected point and the iteration point. These procedures are known as the methods of Uzawa and Rosen.

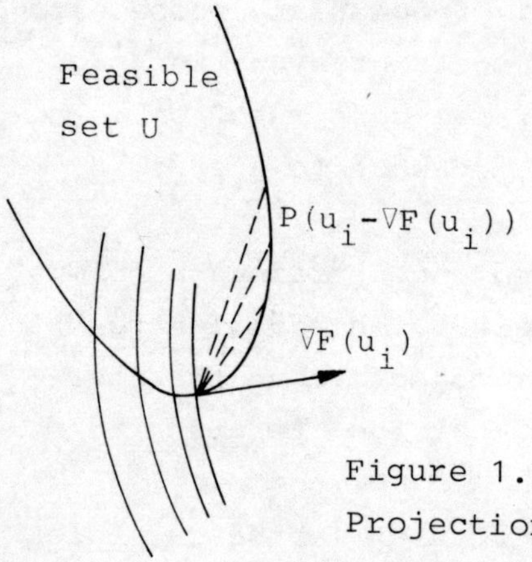


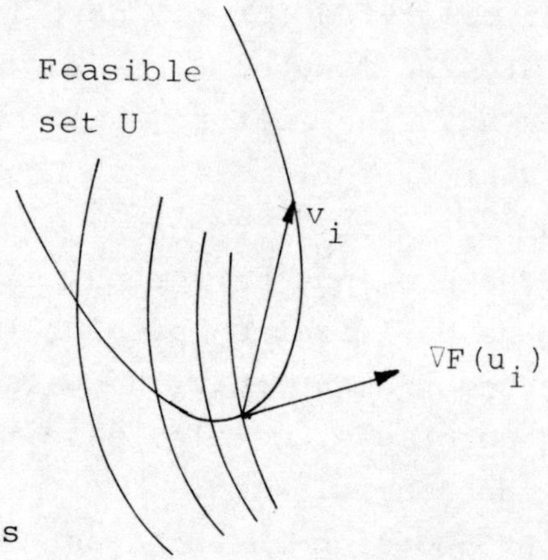


Figure 1.3
Projection methods

## 1.3 OPTIMAL CONTROL OF PARTIAL DIFFERENTIAL EQUATIONS

Consider the problem of a vibrating string which is held fixed at the left endpoint and moved up and down on the right endpoint by the controller:

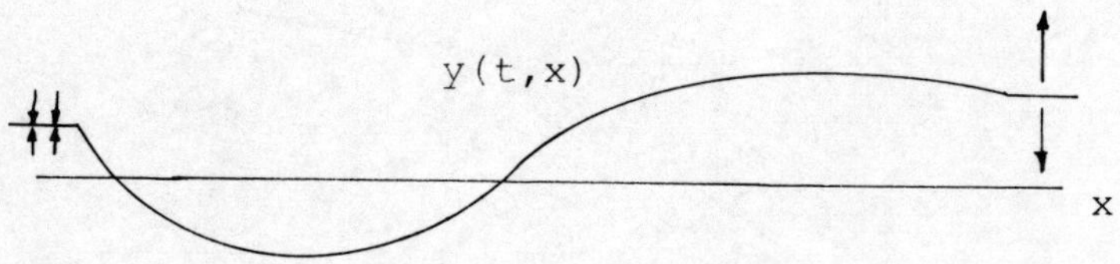


Figure 1.4 Vibrating string

The motion of the string is described by a partial differential equation

$$y_{tt}(t,x) = y_{xx}(t,x) \qquad \text{in} \quad (0,T) \times (0,1)$$

with initial conditions

$$y(0,x) = y_0(x), \quad y_t(0,x) = y_1(x) \qquad \text{in} \quad (0,1)$$

and boundary conditions

$$y(t,0) = 0, \qquad y(t,1) = v(t) \qquad \text{in} \quad (0,T) \; .$$

For each control input v from an approximate space we obtain a unique solution y of the system. The objective function in this case is given by the energy of the vibrating string,

$$\int_0^1 (y_t(T,x)^2 + y_x(T,x)^2) \, dx \; .$$

The numerical solution of this problem is discussed in Chapter 12.

Another class of evolution equations is considered in Chapters 10 and 11. A one dimensional rod is heated or cooled and the temperature is denoted by y(t,x). This diffusion process is governed by

$$y_t(t,x) = y_{xx}(t,x) \qquad \text{in} \quad (0,1) \times (0,T)$$

with an initial temperature distribution

$$y(0,x) = y_0(x) \qquad \text{in} \quad (0,1) \; .$$

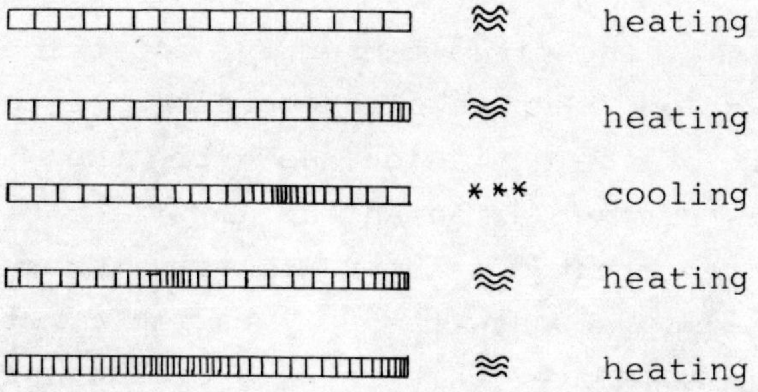


Figure 1.5 Diffusion process

The left endpoint of the rod is insulated , i.e. no heat flow occurs,

$$y_x(t,0) = 0 \quad \text{in} \quad (0,T) .$$

On the right end, the rod is heated or cooled. This process is described by various models.

Using Newton's law on heat transfer we obtain that the temperature gradient, i.e. the rate of the change of the temperature of the medium in the rod, is proportional to the temperature difference between the medium in the rod and the surrounding medium:

$$y_x(t,1) = a(y(t,1) - u(t)) , \tag{1.6}$$

u is the temperature of the controlled medium. If heat transfer occurs by radiation, (1.6) is replaced by

$$y_x(t,1) = b(y(t,1)^4 - u(t)^4) , \tag{1.7}$$

the Stefan-Boltzmann law. Although the control u is nonlinear in (1.7) we can introduce a new control $v(t) = u(t)^4$ making the control input linear. Nevertheless, the nonlinearity of y in (1.7) is substantial and with boundary condition (1.7) we have a nonlinear dependence of the state y on the controls u or v. As the cost functional we select the deviation of the temperature distribution at a certain fixed time $T > 0$ from a given desired curve. The norm can be selected according to the smoothness of the system. In Chapters 10 and 11 a treatment of linear and nonlinear optimal control problems is presented.

Suppose that the final temperature distribution is a continuous function. Then we also could choose the maximum norm as a measure of the deviation. However, then the cost functional lacks Fréchet differentiability. These types of problems have been treated in approximation theory for many years. In Chapter 9 we show how to apply variable metric updates to these problems. Convergence rates are discussed in detail. It is well known that these maximum-norm approximation problems often have a strong uniqueness property. This property is crucial for proofs on convergence rates. In practical applications such as the nonlinear parabolic control problem of the preceding paragraph, the proof of strong uniqueness can be rather deep. A discussion of this property is also contained in Chapter 10.

## 1.4 APPLICATIONS

The last four chapters deal with selected applications of quasi-Newton and projection methods. In the first, we treat an aerospace problem of orbital rendezvous with a moving target in space. As a comet approaches the sun, it passes the earth. The orbit of a satellite is selected in such a way that a close observation of the comet is possible. Given the trajectory of the comet, the orbit of the satellite is changed by impulsive control so that rendezvous is attained. The total fuel used for the maneuver is to be minimized by selecting the magnitude, direction and timing of the impulses.

The second application is an optimal control problem involving a stochastic service system. We consider, as an example, a computer with a job queue waiting to be processed. The queue length is influenced by the service rate of the system, i.e. the number of jobs being completed per time unit, and the arrival rate of the jobs, which changes during the day. The cost functional consists of the cost of providing service and a cost associated with the queue length. An optimal service rate minimizes the sum of these costs.

Chapter 15 treats a problem of production and inventory control. In a deterministic dynamic model of a production system the production rate and workforce level are to be controlled. For given demand, the control inputs of labor, capital and R & D engineering are selected such that the cost of these inputs is minimal.

The last chapter concerns a biomechanical control model describing a worker manually lifting an object. The variables are the angular position of the legs, back and arms, and the forces applied to the different parts of the body. The objective is to minimize the effort expended by the worker during the lifting task such that physical constraints, e.g. limits on the angles and angular rates, and dynamic equations of motion are satisfied.

## 1.5 MEAN VALUE THEOREM

A basic result for convergence proofs is Lagrange's formula or mean value theorem:

THEOREM 1.2: Let U be a convex subset of a normed linear space, $F : U \to \mathbb{R}$ Fréchet differentiable on U. For each $u, v \in U$ there exists $\alpha \in [0,1]$ such that

$$F(u + v) - F(u) = F'_{u+\alpha v}(v) \tag{1.8}$$

or

$$F(u + v) - F(u) = \int_0^1 F'_{u+\eta v}(v)\, d\eta \,. \tag{1.9}$$

If F' is Lipschitz continuous on the line connecting u and u + v then from (1.9)

$$|F(u + v) - F(u) - F'_u(v)| \le \int_0^1 |(F'_{u+\eta v} - F'_u)(v)|\, d\eta$$

$$\le \int_0^1 k\, \eta\, ||v||^2\, d\eta = \frac{k}{2}\, ||v||^2 \,. \tag{1.10}$$

THEOREM 1.3: Let U be a convex set of a normed linear space B, C another normed linear space, $G : U \to C$ Fréchet differentiable on U. For each $u, v \in U$ and each continuous linear functional $\ell$ on C there exists $\alpha \in [0,1]$ such that

$$\ell(G(u + v) - G(u)) = \ell(G'_{u+\alpha v}(v)) \tag{1.11}$$

and

$$\ell(G(u + v) - G(u)) = \int_0^1 \ell(G'_{u+\eta v}(v))\, d\eta \,. \tag{1.12}$$

This implies in turn that for any $g \in G$ there is some $\alpha \in [0,1]$ with

$$||G(u+v) - G(u) - g|| \le ||G'_{u+\alpha v}(v) - g||$$

and

$$||G(u+v) - G(u) - g|| \le \int_0^1 ||G'_{u+\eta v}(v) - g|| \, d\eta \ .$$

For second order analysis we have

<u>THEOREM 1.4:</u> Let U be a convex subset of a normed linear space and $F : U \to \mathbb{R}$ twice Fréchet differentiable on U. For each $u, v \in U$

$$F(u+v) = F(u) + F'_u(v) + \int_0^1 \int_0^1 \eta \, F''_{u+\sigma\eta v}(v)(v) d\eta d\sigma \ . \qquad (1.13)$$

By Theorem 1.2

$$F(u+v) - F(u) - F'_u(v) = \int_0^1 (F'_{u+\eta v} - F'_u)(v) d\eta \ .$$

By assumption the map

$$\eta \to F'_{u+\eta v}(v)$$

is differenentiable and (1.13) follows.

(1.13) implies that if for $u \in U$, $v \in U$

$$m \ ||v||^2 \le F''_u(v)(v) \le M \ ||v||^2$$

then

$$\frac{m}{2} \ ||v||^2 \le F(u+v) - F(u) - F'_u(v) \le \frac{M}{2} \ ||v||^2 \ . \qquad (1.14)$$

<u>THEOREM 1.5:</u> Let F be twice Fréchet differentiable on a convex set $U \subset H$. If there are reals $m, M > 0$ such that

$$m \langle u,u\rangle \le \langle u, \nabla^2 F(v)(u)\rangle \le M \langle u,u\rangle \qquad (1.15)$$

for all $v \in U$, $u \in H$, then for each $u_1, u_2 \in U$ with

$$p = u_1 - u_2$$

and

$$y = \nabla F(u_1) - \nabla F(u_2) \ ,$$

we have the following inequalities:

$$m \langle p,p\rangle \le \langle p,y\rangle \le M \langle p,p\rangle \,, \tag{1.16}$$

$$m^2 \langle p,p\rangle \le \langle y,y\rangle \le M^2 \langle p,p\rangle \,. \tag{1.17}$$

Proof: Using (1.9)

$$\langle p,y\rangle = \langle u_1 - u_2, \nabla F(u_1) - \nabla F(u_2)\rangle$$

$$= \int_0^1 \langle u_1 - u_2, \nabla^2 F(u_2 + \eta(u_1 - u_2))(u_1 - u_2)\rangle \, d\eta$$

yielding (1.16).

Condition (1.15) implies the Lipschitz continuity of $\nabla F$ on U,

$$\langle y,y\rangle = \|\nabla F(u_1) - \nabla F(u_2)\|^2 \le M^2 \|u_1 - u_2\|^2 \,.$$

With (1.16) we obtain

$$m^2 \langle p,p\rangle^2 \le \langle p,y\rangle^2 \le \langle p,p\rangle \langle y,y\rangle$$

which implies (1.17).

# 2 Step length determination

## 2.1 ZOUTENDIJK'S CONVERGENCE CONDITION

In many optimization algorithms not only the descent direction but also the step length in that direction must be determined. We shall consider step length rules for which we obtain a convergence proof via Zoutendijk's condition.

We impose the following

ASSUMPTION $A_1$: Let E be a Banach space and $F : E \to \mathbb{R}$ be Fréchet differentiable with Lipschitz continuous derivative F' on the bounded set

$$S := \{u \in E : F(u) \leq F(u_o)\} ,$$

for some starting point $u_o \in E$. Let $\{u_i\}_{\mathbb{N}}$ be a sequence of iterates in E and $\{v_i\}_{\mathbb{N}}$ a sequence of descent directions, i.e.

$$F'_{u_i}(v_i) < 0 \qquad \text{for all } i \in \mathbb{N},\ v_i \in E,$$

where $u_{i+1}$ is determined by

$$u_{i+1} = u_i + \alpha_i v_i \tag{2.1}$$

with a certain step length rule for $\alpha_i \in \mathbb{R}$ .

In the following sections we shall prove for several step length rules that the

$$\underline{\text{Zoutendijk condition}} \qquad \sum_{i=o}^{\infty} (F'_{u_i}(v_i)\, \|v_i\|^{-1})^2 < \infty \tag{2.2}$$

is fulfilled. The use of this condition for convergence proofs will be encountered in the sequel.

## 2.2 EXACT STEP LENGTH RULE

THEOREM 2.1: Let $(A_1)$ be satisfied and the step length be determined by the first local extremum, i.e.

$$\alpha_i = \min \{\alpha > 0 : F'_{u_i + \alpha v_i}(v_i) = 0\} . \tag{2.3}$$

Then the Zoutendijk condition (2.2) is fulfilled.

Proof: The number $\alpha_i$ defined in (2.3) exists because of $(A_1)$. Using the Lipschitz continuity of F' with constant k we obtain from (2.3) inserting $F'_{u_i}(v_i)$,

$$0 = F'_{u_i + \alpha_i v_i}(v_i) \leq F'_{u_i}(v_i) + k\alpha_i \|v_i\|^2$$

and hence

$$\alpha_i \geq - F'_{u_i}(v_i) \; (k \|v_i\|^2)^{-1} =: \mu_i \; . \tag{2.4}$$

By the mean value theorem in Section 1.5 and (2.4)

$$F(u_{i+1}) - F(u_i) \leq \int_0^{\mu_i} F'_{u_i+\sigma v_i}(v_i)d\sigma$$

$$\leq \int_0^{\mu_i} (F'_{u_i}(v_i) + k\sigma \|v_i\|^2)d\sigma = -\frac{1}{2k} F'_{u_i}(v_i)^2 \|v_i\|^{-2} \; .$$

This implies

$$F(u_{i+1}) - F(u_o) = \sum_{j=0}^{i} (F(u_{j+1}) - F(u_j))$$

$$\leq - \sum_{j=0}^{i} \frac{1}{2k} (F'_{u_j}(v_j))^2 \|v_j\|^{-2}$$

and by the boundedness of F on S for $i \to \infty$ we see that (2.2) holds.

## 2.3 ARMIJO'S STEP LENGTH RULE

In the following we use a stronger assumption than $(A_1)$:

ASSUMPTION $A_2$: Let $(A_1)$ be satisfied and F be twice Fréchet differentiable such that there are $m, M > 0$ with

$$m \|u\|^2 \leq F''_v(u)(u) \leq M \|u\|^2 \quad \text{for all } u \in E, \; v \in S \; .$$

THEOREM 2.2: Let $(A_2)$ be satisfied and $\gamma \in (0,1)$ given. The step length $\alpha_i = \gamma^{k(i)}$ is determined by the smallest integer $k(i)$ that fulfills the inequalities

$$F(u_i + \alpha_i v_i) - F(u_i) \leq \gamma\alpha_i F'_{u_i}(v_i) \; , \tag{2.5}$$

$$F(u_i + \gamma^{-1}\alpha_i v_i) - F(u_i) \geq \alpha_i F'_{u_i}(v_i) \; . \tag{2.6}$$

Then the Zoutendijk condition (2.2) is true.

Proof: $(A_2)$ implies the existence of k(i). Since F is bounded from below by $(A_2)$ we conclude from (2.5)

$$- \infty < \sum_{i=0}^{\infty} \alpha_i F'_{u_i}(v_i) < 0 .$$

We claim that for all $i \in \mathbb{N}$

$$\alpha_i \geq - \gamma F'_{u_i}(v_i) \quad (M \|v_i\|^2)^{-1} (1 - \gamma), \tag{2.7}$$

which would yield (2.2).

If (2.7) were not true for some $j \in \mathbb{N}$, then

$$\alpha_j < - \gamma F'_{u_i}(v_i) \quad (M \|v_j\|^2)^{-1} (1 - \gamma) \tag{2.8}$$

and by the mean value theorem in Section 1.5 and (2.8)

$$F(u_j + \gamma^{-1} \alpha_j v_j) - F(u_j) \leq \gamma^{-1} (\alpha_j F'_{u_j}(v_j) + M\alpha_j^2 \|v_j\|^2$$

$$< \alpha_j F'_{u_j}(v_j).$$

Hence (2.5) is satisfied for $\bar{\alpha}_j = \gamma^{k(i)-1}$. If (2.6) is also true for $\bar{\alpha}_j$, then we have a contradiction, since k(i) is not the smallest integer that fulfills (2.5), (2.6). If (2.6) is not true for $\bar{\alpha}_j$, then redefine $\bar{\alpha}_j = \gamma^{k(i)-2}$ and (2.5) is satisfied for this $\bar{\alpha}_j$. Then the argument is repeated. Applying this procedure iteratively, there exists an $\bar{\alpha}_j$ such that (2.6) holds, because otherwise F would be unbounded from below, which is not true from $(A_2)$.

## 2.4 GOLDSTEIN'S STEP LENGTH RULE

THEOREM 2.3: Let $(A_2)$ be satisfied and $\gamma \in (0,0.5)$ given. The step length $\alpha_i > 0$ fulfills

$$\alpha_i (1-\gamma) F'_{u_i}(v_i) \leq F(u_i + \alpha_i v_i) - F(u_i) \leq \alpha_i \gamma F'_{u_i}(v_i). \tag{2.9}$$

Then the Zoutendijk condition (2.2) holds.

Proof: The mean value theorem in Section 1.5 gives

$$F(u_i + \alpha_i v_i) - F(u_i)$$

$$= \alpha_i F'_{u_i}(v_i) + \alpha_i^2 \int_0^1 \int_0^1 F''_{u_i + \sigma\eta\alpha_i v_i}(v_i)(v_i) \eta \, d\sigma \, d\eta .$$

This implies using $(A_2)$

$$\alpha_i F'_{u_i}(v_i) + \alpha_i^2 m \|v_i\|^2/2 \leq F(u_i + \alpha_i v_i) - F(u_i)$$

$$\leq \alpha_i F'_{u_i}(v_i) + \alpha_i^2 M \|v_i\|^2/2 . \tag{2.10}$$

Combining (2.10) and (2.9) we obtain

$$\alpha_i (1-\gamma) F'_{u_i}(v_i) \leq \alpha_i F'_{u_i}(v_i) + \alpha_i^2 M \|v_i\|^2/2$$

and

$$-2\gamma F'_{u_i}(v_i) / (M \|v_i\|^2) \leq \alpha_i . \tag{2.11}$$

Boundedness of F yields by (2.9)

$$-\infty < \sum_{i=0}^{\infty} \alpha_i F'_{u_i}(v_i) < O$$

and with (2.11) the Zoutendijk condition (2.2).

## 2.5 POWELL'S STEP LENGTH RULE

THEOREM 2.4: Let $(A_2)$ be satisfied and $\gamma, \mu \in \mathbb{R}$ with $O < \gamma \leq \mu < 1$ be given. The step length $\alpha_i > O$ may fulfill the inequalities

$$F(u_i + \alpha_i v_i) - F(u_i) \leq \gamma \alpha_i F'_{u_i}(v_i) \tag{2.12}$$

$$\mu F'_{u_i}(v_i) \leq F'_{u_i+\alpha_i v_i}(v_i). \tag{2.13}$$

Then the Zoutendijk condition (2.2) holds.

Proof: From (2.13) we conclude via the mean value theorem

$$(\mu - 1) F'_{u_i}(v_i) \leq \alpha_i \int_0^1 F''_{u_i+\eta\alpha_i v_i}(v_i)(v_i)d\eta \leq \alpha_i M \|v_i\|^2$$

which implies

$$-(1-\mu) F'_{u_i}(v_i) / (M \|v_i\|^2) \leq \alpha_i . \tag{2.14}$$

(2.12) leads to

$$-\infty < \sum_{i=0}^{\infty} \alpha_i F'_{u_i}(v_i) < O ,$$

which yields with (2.14) the Zoutendijk condition (2.2).

## 2.6 UNDERESTIMATION

In contrast to the step length rule in 2.2, all the other rules in 2.3 - 2.5 are implemented by a finite procedure, since only some point in a prescribed interval must be determined. Sometimes it is possible to indicate whether these intervals contain the optimal step size or whether it lies on the left or right of the interval. The latter case is called an under- or overestimation of the step size. There are controversial opinions concerning which cases are desirable, see e.g. [ 1, p.314] and [ 2, p.1248]. We limit this treatment to some theoretical results on estimation.

LEMMA 2.5: Let $(A_2)$ hold and $u_i, v_i \in H$ such that

$$F'_{u_i}(v_i) < 0 .$$

Then the optimal step size $\alpha_i^*$ defined in (2.3) satisfies

$$\alpha_i^* \in [M^{-1}\,\delta_i,\ m^{-1}\,\delta_i] \tag{2.15}$$

with

$$\delta_i = -F'_{u_i}(v_i)\ \|v_i\|^{-2} . \tag{2.16}$$

Proof: Using the mean value theorem,

$$0 = F'_{u_i+\alpha_i^* v_i}(v_i) = \alpha_i^* \int_0^1 F''_{u_i+\eta\alpha_i^* v_i}(v_i)(v_i)d\eta + F'_{u_i}(v_i)$$

which implies with $(A_2)$ the estimate (2.15).

THEOREM 2.6: Let $(A_2)$ hold and $\alpha_i$ be determined by Armijo's, Goldstein's or Powell's step length rule. If

$$\gamma \geq 1 - \frac{m}{2M} \tag{2.17}$$

with $\gamma$ in (2.5), (2.9) or (2.12), then $\alpha_i$ always underestimates the optimal step length.

Proof: In all quoted step length rules we have

$$F(u_i + \alpha_i v_i) - F(u_i) \leq \gamma\,\alpha_i\,F'_{u_i}(v_i) . \tag{2.18}$$

We claim that

$$\alpha_i \leq 2\delta_i\,\frac{1-\gamma}{m} , \tag{2.19}$$

for, if this were not true, we obtain from (2.10)

$$F(u_i + \alpha_i v_i) - F(u_i) \geq m\alpha_i^2 \|v_i\|^2 / 2 + \alpha_i F'_{u_i}(v_i)$$

$$> \gamma \alpha_i F'_{u_i}(v_i) ,$$

a contradiction to (2.18).

From (2.17) and (2.19) we deduce

$$\alpha_i \leq 2\delta_i \frac{1-\gamma}{m} \leq \delta_i M^{-1} ,$$

which implies with Lemma 2.5

$$\alpha_i \leq \alpha_i^* .$$

## 2.7 COMMENTS

Section 2.1: The convergence condition (2.2) is given by Zoutendijk in [3].

Section 2.2: Sometimes this rule is credited to Curry [4].

Section 2.3: This step size rule was presented in [5].

Section 2.4: Goldstein [6] developed this step length rule.

Section 2.5: In connection with variable metric methods Powell [7] introduced this step size rule.

Section 2.6: Depending on the profile of the objective functional it is sometimes desirable to have an inexact step length rule which yields step sizes that are always smaller or larger than optimal.

In the literature there are several other successful approaches for unifying convergence theories; see Daniel [8], Polak [9] or Zangwill [10].

# 3 Variable metric updates

## 3.1 MOTIVATION

Let F be a real valued function defined on a Hilbert space H, which is twice Fréchet differentiable. We consider the following unconstrained optimization problem:

Find $\hat{u} \in H$ such that

$$F(\hat{u}) \leq F(u) \quad \text{for all } u \in H . \tag{3.1}$$

From an algorithmic point of view, the problem is to find with a given approximation $u_i \in H$ of an optimal point, a "better" approximation $u_{i+1} \in H$. A common procedure is to replace the function F by first or second order approximations and to solve (3.1) for the truncated F. First order approximations lead to usual gradient methods whereas the second order approximation gives Newton's method as follows.

Taylor-expansion $\tilde{F}$ of F at $u_i$

$$F(u) \approx \tilde{F}(u) = F(u_i) + < \nabla F(u_i), u - u_i > +$$
$$+ \frac{1}{2} < u - u_i, \nabla^2 F(u_i)(u - u_i) >$$

can be used in (3.1) to obtain the approximating problem:

Find $u_{i+1} \in H$ such that

$$\tilde{F}(u_{i+1}) \leq \tilde{F}(u) \quad \text{for all } u \in H . \tag{3.2}$$

Problem (3.2) is equivalent to the statement that the gradient of $\tilde{F}$ vanishes at $u_{i+1}$, i.e.

$$\Theta = \nabla \tilde{F}(u_{i+1}) = \nabla F(u_i) + \nabla^2 F(u_i)(u_{i+1} - u_i) .$$

Assuming invertibility of $\nabla^2 F(u_i)$, we obtain

$$u_{i+1} = u_i - (\nabla^2 F(u_i))^{-1} \nabla F(u_i) ,$$

which is Newton's method for problem (3.1). It is well known that this method has very good local convergence properties. However, in various applications the overall computational effort of evaluating second derivatives is high. The disadvantage is avoided by variable metric or quasi-Newton methods in which the second derivative operator is replaced by an operator $H_i$ which is updated at each iteration only using first order derivatives. Various properties which seem desirable for defining the updates will be treated in the following sections. The prototype of a variable metric algorithm is as follows:

0. Select $u_o \in H$, a linear invertible bounded operator $B_o : H \to H$, a step length rule and an update rule for $B_i$.
1. Compute $\nabla F(u_i)$ and $v_i = -B_i^{-1} \nabla F(u_i)$.
2. Find $\alpha_i \in (o, \infty)$ by the step length rule.
3. Set $u_{i+1} = u_i + \alpha_i v_i$.
4. Compute $B_{i+1}$ by the update rule.
5. Set $i + 1 = i$ and go to 1.

Termination is achieved if, for example, $\|\nabla F(u_i)\|$ is less than a certain small number.

## 3.2 RANK-ONE AND RANK-TWO UPDATES

Although there are various rules for updating $B_i$, the most successful has been to add a correction term $T_i$ such that

$$B_{i+1} = B_i + T_i .$$

In order that T has a rather simple form, we shall assume that

$$T \in L(H) = \{T : H \to H : T \text{ linear, continuous}\}$$

and

$$\dim TH \le 2 .$$

Let $c, d \in H$ such that $\text{span}\{c,d\} = TH$, i.e. for each $u \in H$ there are $\zeta(u), \delta(u) \in \mathbb{R}$ with

$$T u = \zeta(u)c + \delta(u)d \quad \text{for all } u \in H. \tag{3.3}$$

Since $T \in L(H)$, we obtain from (3.3) that $\zeta, \delta \in H^*$, the topological dual space of H. Hence $\zeta, \delta$ are represented by $a, b \in H$ such that (3.3) can be written,

$$T\,u = \langle a,u\rangle\, c + \langle b,u\rangle\, d \;. \tag{3.4}$$

This relation gives rise to use the notation of dyadic product:

DEFINITION 3.1: $v,w \in H$ define an operator

$$]v,w[\; \in L(H)$$

by

$$]v,w[\;(u) = \langle w,u\rangle\, v \qquad \forall u \in H \;.$$

We recall some formulas for these operators:
For any $A \in L(H)$

$$A\,]v,w[\; = \;]Av,w[\;, \tag{3.5}$$

$$]v_1,w_1[\;]v_2,w_2[\; = \langle w_1,v_2\rangle\; ]v_1,w_2[\;. \tag{3.6}$$

Hence we can conclude this section with

LEMMA 3.2: Each $T \in L(H)$ with $\dim TH \le 2$ is represented by $a_i \in H$, $i = 1,\dots,4$ and

$$T = \;]a_1,a_2[\; + \;]a_3,a_4[\;. \tag{3.7}$$

## 3.3 INVERTIBILITY OF THE UPDATES

From step 1 of the prototype algorithm in 3.1 it is intuitively clear that each operator of the sequence $\{B_i\}_{\mathbb{N}}$ has to be invertible. Therefore, we prove the following theorem:

THEOREM 3.3: Let B be an automorphism on H and $a_i \in H$, $i = 1,..,4$. Then

$$\bar{B} = B + \;]a_1,a_2[\; + \;]a_3,a_4[$$

is an automorphism on H if and only if

$$\alpha = (1 + \langle a_2, B^{-1}a_1\rangle)\,(1 + \langle a_4, B^{-1}a_3\rangle) - \langle a_2, B^{-1}a_3\rangle\,\langle a_4, B^{-1}a_1\rangle \neq 0\;, \tag{3.8}$$

and the inverse of $\bar{B}$ is given by

$$\bar{B}^{-1} = B^{-1} - \alpha^{-1}B^{-1}\,((1 + \langle a_4, B^{-1}a_3\rangle)\,]a_1,a_2[\; - \langle a_2, B^{-1}a_3\rangle\; ]a_1,a_4[\; - (1 + \langle a_2, B^{-1}a_1\rangle)\; ]a_3,a_4[)\,B^{-1}$$

$$- \langle a_4, B^{-1} a_1 \rangle \, ]a_3, a_2[ \,) \, B^{-1} \, .$$

Proof: Let $v \in H$ be given. Then there exists a unique $u \in H$ with $\bar{B}u = v$ if and only if the equation

$$u = - \langle a_2, u \rangle \, B^{-1} a_1 - \langle a_4, u \rangle \, B^{-1} a_3 + B^{-1} v \tag{3.9}$$

is uniquely solvable for $u \in H$. Let us assume that $a_1$ and $a_3$ are linearly independent, otherwise we can set $a_3 = \Theta$ and the proof is similar for this technically less complicated case. By (3.9) u is of the form

$$u = \beta B^{-1} a_1 + \gamma B^{-1} a_3 + \nu B^{-1} v \, .$$

Inserting u into (3.9) and comparing coefficients, (3.9) is uniquely solvable in the general case (i.e. $a_1, a_3$ and v are linearly independent) if and only if the following system of equations is uniquely solvable for $\beta, \gamma, \nu \in \mathbb{R}$ for all $v \in H$:

$$\beta = - \langle a_2, u \rangle = - \beta \langle a_2, B^{-1} a_1 \rangle - \gamma \langle a_2, B^{-1} a_3 \rangle - \nu \langle a_2, B^{-1} v \rangle$$

$$\gamma = - \langle a_4, u \rangle = - \beta \langle a_4, B^{-1} a_1 \rangle - \gamma \langle a_4, B^{-1} a_3 \rangle - \nu \langle a_4, B^{-1} v \rangle$$

$$\nu = 1 \, .$$

This holds if and only if the determinant of the coefficient matrix does not vanish which is identical with (3.8).

The representation of $\bar{B}^{-1}$ follows by inspection.

For the case of rank-one updates Theorem 3.3 simplifies considerably.

COROLLARY 3.4: Let B be an automorphism on H and $v, w \in H$. Then $\bar{B} = B + ]v,w[$ is an automorphism if and only if

$$\alpha = 1 + \langle w, B^{-1} v \rangle \neq 0$$

and in this case

$$\bar{B}^{-1} = B^{-1} - \alpha^{-1} B^{-1} \, ]v,w[ \, B^{-1} \, .$$

## 3.4 SELF-ADJOINT UPDATES

The second order derivative operator is self-adjoint. Hence it is reasonable to require this property also for its approximations $\{B_i\}_{\mathbb{N}}$. Starting with a self-adjoint $B_o$ we seek those updates $T_i$ such that, if $B_i$ is self-adjoint, $B_i + T_i$ also has this property. This holds if and only if $T_i$ is self-adjoint. Therefore we introduce the following linear subspace of $L(H)$ for $a,c \in H$.

$$S(a,c) = \{T \in L(H) \mid TH = \text{span}\{a,c\},\ T \text{ self-adj.}\}\ .$$

<u>THEOREM 3.5:</u> Each $T \in L(H)$ with $\dim TH \leq 2$ is self-adjoint if and only if there are $a,c \in H$, $\alpha,\beta,\gamma \in \mathbb{R}$ such that $T$ is represented by

$$T = \alpha\ ]a,a[\ + \beta\ ]c,c[\ + \gamma\ (]a,c[\ + ]c,a[)\ . \qquad (3.10)$$

<u>Proof:</u> We prove that $S(a,c)$ is identical with

$$S_1 = \{T \in L(H) \mid T \text{ represented by (3.10)},\ \alpha,\beta,\gamma \in \mathbb{R}\}\ .$$

Case 1: $a,c$ linearly dependent. Then $S_1$ reduces to

$$S_1 = \{T \in L(H) \mid T = \alpha\ ]a,a[,\ \ \alpha \in \mathbb{R}\}.$$

Let $T \in S(a,c)$, i.e. $T = ]a,b[$ for some $b \in H$. $T$ self-adjoint implies

$$\langle a,v\rangle\ \langle b,w\rangle = \langle a,w\rangle\ \langle b,v\rangle \quad \text{for all } w,v \in H$$

$$\Leftrightarrow\ \ \langle a,v\rangle b = \langle b,v\rangle a \qquad \text{for all } v \in H$$

$$\Leftrightarrow\ \ b = \alpha a \Rightarrow T = \alpha\ ]a,a[\ \in S_1\ .$$

The reverse inclusion is trivial.

Case 2: $a,c$ linearly independent.
$S(a,c) \supset S_1$ is trivial. Furthermore $S_1$ contains three linearly independent operators $]a,a[$, $]c,c[$, $]a,c[\ + ]c,a[$, because, if there would exist $\alpha,\beta,\gamma \in \mathbb{R}$ such that

$$\alpha\ ]a,a[\ + \beta\ ]c,c[\ + \gamma\ (]a,c[\ + ]c,a[) = \Theta\ ,$$

then for all $u \in H$

$$\alpha \langle a,u\rangle + \gamma \langle c,u\rangle = 0 \quad \text{and} \quad \beta \langle c,u\rangle + \gamma \langle a,u\rangle = 0$$

and hence

$$\alpha a + \gamma c = \Theta \quad \text{and} \quad \beta c + \gamma a = \Theta$$

by linear independence

$$\alpha = \beta = \gamma = 0 .$$

This proves $\dim S_1 = 3$.

It remains to show that $\dim S(a,c) \le 3$. Assume, that $\dim S(a,c) > 3$ is true. Then there exist $T_1,\dots,T_4 \in S(a,c)$

$$T_i = ]a,b_i[ + ]c,d_i[ \ , \quad b_i,\ d_i \in H \ , \quad i = 1,\dots,4 \ ,$$

which are linearly independent. $T_i$ self-adjoint gives

$$\langle a,u\rangle \langle b_i,v\rangle + \langle c,u\rangle \langle d_i,v\rangle$$

$$= \langle a,v\rangle \langle b_i,u\rangle + \langle c,v\rangle \langle d_i,u\rangle$$

for all $u,v \in H$. From this we deduce

$$\langle a,u\rangle b_i + \langle c,u\rangle d_i = \langle b_i,u\rangle a + \langle d_i,u\rangle c \tag{3.11}$$

for all $u \in H$.

If for some $1 \le i \le 4$ the set of vectors $\{a,c,b_i,d_i\}$ is linearly independent, we obtain from (3.11)

$$a = b_i = c = d_i = 0 \ ,$$

a contradiction.

Hence, let for some $1 \le i \le 4$ the set $\{a,c,b_i\}$ consist of linearly independent vectors. From (3.11) follows for all $u \in H$ with $\langle c,u\rangle \neq 0$

$$d_i = \langle c,u\rangle^{-1} (\langle b_i,u\rangle a - \langle a,u\rangle b_i + \langle d_i,u\rangle c) \tag{3.12}$$

and for all $u \in H$ with $\langle c,u\rangle = 0$

$$\langle b_i,u\rangle = \langle a,u\rangle = \langle d_i,u\rangle = 0 .$$

By linear independence the coefficient of a in (3.12) is uniquely determined, say $\alpha$, and hence

$$\langle b_i,u\rangle = \alpha \langle c,u\rangle \qquad \text{for all } u \in H .$$

This implies $b_i = \alpha c$, a contradiction to the linear independence.

Hence, each set $\{a,b_i,c,d_i\}$ , $i = 1,\dots,4$, contains at most two linearly independent vectors. Since a,c are linearly independent by assumption, there exist $\alpha_i, \bar{\alpha}_i, \zeta_i, \bar{\zeta}_i \in \mathbb{R}$ for $i = 1,\dots,4$ with

$$b_i = \alpha_i a + \zeta_i c , \quad d_i = \bar{\alpha}_i a + \bar{\zeta}_i c , \qquad i = 1,\dots,4. \quad (3.13)$$

Substituting (3.13) into the left side of (3.11)

$$(\alpha_i \langle a,u\rangle + \bar{\alpha}_i \langle c,u\rangle)a + (\zeta_i \langle a,u\rangle + \bar{\zeta}_i \langle c,u\rangle)c$$
$$= \langle b_i,u\rangle a + \langle d_i,u\rangle c$$

for all $u \in H$ implies

$$b_i = \alpha_i a + \bar{\alpha}_i c , \quad d_i = \zeta_i a + \bar{\zeta}_i c , \qquad i = 1,\dots,4$$

and with $b_i$ by (3.13)

$$b_i = \alpha_i a + \zeta_i c , \quad d_i = \zeta_i a + \bar{\zeta}_i c , \qquad i = 1,\dots,4. \quad (3.14)$$

For given $\alpha_i, \zeta_i, \bar{\zeta}_i$ , $i = 1,\dots,4$, there exist $\gamma_i \in \mathbb{R}$, $i = 1,\dots,4$, not all zero, such that

$$\sum_{i=1}^{4} \gamma_i \alpha_i = \sum_{i=1}^{4} \gamma_i \zeta_i = \sum_{i=1}^{4} \gamma_i \bar{\zeta}_i = 0 .$$

In connection with (3.14), this implies

$$\sum_{i=1}^{4} \gamma_i b_i = \sum_{i=1}^{4} \gamma_i d_i = \Theta$$

and

$$\sum_{i=1}^{4} \gamma_i T_i u = \Theta \quad \text{for all } u \in H ,$$

a contradiction to the linear independence of $\{T_i\}_{i=1}^4$ . Hence $\dim S(a,c) \le 3$ which completes the proof.

## 3.5 UPDATES SATISFYING AN EQUATION

After the i-th iteration we have $u_i$, $u_{i+1}$, $\nabla F(u_i)$, $\nabla F(u_{i+1})$ available for constructing $B_{i+1}$. Expanding $\nabla F(u_{i+1})$ at $u_{i+1}$ we obtain

$$\nabla F(u_{i+1}) \simeq \nabla F(u_i) + \nabla^2 F(u_{i+1})\,(u_{i+1} - u_i) .$$

Since $B_{i+1}$ is used as an approximation of $\nabla^2 F(u_{i+1})$, it is useful to impose the following condition on $\{B_i\}_{\mathbb{N}}$ :

Each $B_{i+1}$, $i \in \mathbb{N}_o$, satisfies the quasi-Newton equation

$$B_{i+1}(u_{i+1} - u_i) = \nabla F(u_{i+1}) - \nabla F(u_i) . \qquad (3.15)$$

Hence we consider in this section updates which satisfy an inhomogeneous equation.

LEMMA 3.6: Let $p,z \in H$ be given. Then the set $R(p,z)$ of all $T \in L(H)$ such that $T$ is self-adjoint, $\dim TH \le 2$ and $Tp = z$ is given by

$$R(p,z) = \{T = \alpha\, ]a,a[ + \beta\, ]z,z[ + \gamma\, (]a,z[ + ]z,a[) \mid$$
$$a \in H,\ \alpha,\beta,\gamma \in \mathbb{R} \quad \text{such that}$$
$$\alpha \langle a,p\rangle + \gamma \langle z,p\rangle = 0 \quad \text{and}$$
$$\beta \langle z,p\rangle + \gamma \langle a,p\rangle = 1 \} . \qquad (3.16)$$

Proof: By inspection, each $T \in R(p,z)$ satisfies $Tp = z$ and $T \in S(a,z)$. On the other hand, let $T \in L(H)$ be self-adjoint, $Tp = z$ and $\dim TH \le 2$. Then $TH = \operatorname{span}\{a,c\}$ for some $a,c \in H$ and, since $z \in TH$, there are $\alpha,\zeta \in \mathbb{R}$ with $z = \alpha a + \zeta c$. Assume for definiteness $\zeta \neq 0$. Then $\operatorname{span}\{a,c\} = \operatorname{span}\{a,z\}$ and $T \in S(a,z)$. Using the representation (3.10) we obtain

$$z = Tp = (\alpha \langle a,p\rangle + \gamma \langle z,p\rangle)a + (\beta \langle z,p\rangle + \gamma \langle a,p\rangle)z$$

which holds if and only if $T \in R(p,z)$.

We derive a different representation for $R(p,z)$.

THEOREM 3.7: For $p,z \in H$, $p,z \neq \Theta$, we have

$$R(p,z) = \{T = (\zeta \langle z,p\rangle + \langle b,p\rangle^2)^{-1} (\zeta\, ]z,z[$$
$$- \langle z,p\rangle\, ]b,b[ + \langle b,p\rangle\, (]b,z[ + ]z,b[)):$$
$$\zeta \in \mathbb{R},\ b \in H,\ \zeta \langle z,p\rangle + \langle b,p\rangle^2 \neq 0\} . \qquad (3.17)$$

Proof: It is easy to check that any $T$ represented by (3.17) lies in $R(p,z)$.

Let $T = \alpha\, ]a,a[ + \beta\, ]z,z[ + \gamma(]a,z[ + ]z,a[)$

with (3.16) be given. We consider two cases:

$1 - \beta \langle z,p\rangle \neq 0$. Then by (3.16) $\langle a,p\rangle \neq 0$

and

$$\gamma = (1 - \beta \langle z,p\rangle) \langle a,p\rangle^{-1} ,$$

$$\alpha = - \langle z,p\rangle (1 - \beta \langle z,p\rangle) \langle a,p\rangle^{-2} .$$

Define b, $\zeta$ by

$$b = a \quad \text{und} \quad \zeta = \beta \langle b,p \rangle^2 (1 - \beta \langle z,p\rangle)^{-1} .$$

We obtain

$$\zeta \langle z,p\rangle + \langle b,p\rangle^2 = \langle b,p\rangle^2 (1 - \beta \langle z,p\rangle)^{-1} = \zeta\beta^{-1}$$

(for $\beta = 0$ it is evident that R is of type (3.17) ) and

$$\beta = \zeta (\zeta \langle z,p\rangle + \langle b,p\rangle^2)^{-1} ,$$

$$\gamma = \langle b,p\rangle (\zeta \langle z,p\rangle + \langle b,p\rangle^2)^{-1} ,$$

$$\alpha = - \langle z,p\rangle (\zeta \langle z,p\rangle + \langle b,p\rangle^2)^{-1} .$$

Hence T is of the type given in (3.17).

$1 - \beta \langle z,p\rangle = 0$. Then $\langle z,p\rangle \neq 0$ and by (3.16) $\gamma \langle a,p\rangle = 0$. If $\langle a,p\rangle \neq 0$, then, from (3.16), $\gamma = \alpha = 0$ and $T = \beta\, ]z,z[ = \langle z,p\rangle^{-1}\, ]z,z[$, which is representable in the form of (3.17) by choosing $b = \Theta$, $\zeta \neq 0$. Omitting this trivial case, we assume $\langle a,p\rangle = 0$. Then by (3.16) $\gamma = 0$ and

$$T = \alpha\, ]a,a[ + \langle z,p\rangle^{-1}\, ]z,z[ . \qquad (3.18)$$

For $\alpha = 0$ we have the same situation as above. If $\alpha \neq 0$, then choose $b = a$ and $\zeta = \alpha^{-1}$. Using the assumption $\langle b,p\rangle = \langle a,p\rangle = 0$ T is written as in (3.17).

In the following lines we give some examples of known updates. However, first we have to interpret the vector z. Defining

$$p_i = u_{i+1} - u_i , \quad y_i = \nabla F(u_{i+1}) - \nabla F(u_i) , \qquad (3.19)$$

the quasi-Newton equation (3.15) is $B_{i+1}p_i = y_i$ or by omitting indices and setting $\overline{B} = B_{i+1}$

$$\overline{B}p = Bp + Tp = y .$$

Hence $z = y - Bp$. In the past five years, so called "scaled" versions of quasi-Newton methods have been developed. Instead of updating $\overline{B}$ by $\overline{B} = B + T$, we employ

$$\overline{B} = \rho B + T \, . \tag{3.20}$$

The new parameter $\rho$ is determined at each iteration. In this case, we have

$$z = y - \rho Bp \, . \tag{3.21}$$

Since we have not yet made use of the special structure of z, we can define classes of updates which can also be used for scaling methods.

Choosing $\zeta = 0$ in (3.17) we obtain a scaled version of the class of methods considered in $\mathbb{R}^n$ by Dennis-Moré:

$$\overline{B} = \rho \, (B + \frac{\langle p,Bp\rangle}{\langle b,p\rangle^2} \, ]b,b[ \, - \frac{1}{\langle b,p\rangle} \, (]Bp,b[ \, + \, ]b,Bp[))$$

$$- \frac{\langle y,p\rangle}{\langle b,p\rangle^2} \, ]b,b[ \, + \frac{1}{\langle b,p\rangle} \, (]y,b[ \, + \, ]b,y[) \, . \tag{3.22}$$

For b = p we obtain a scaled version of Powell's symmetric Broyden update (PSB update)

$$\overline{B} = \frac{\rho}{\langle p,p\rangle^2} \, (\langle p,p\rangle^2 B + \langle p,Bp\rangle \, ]p,p[ \, - \langle p,p\rangle \, (]Bp,p[ \, + \, ]p,Bp[)$$

$$+ \frac{1}{\langle p,p\rangle^2} \, (\langle p,p\rangle \, (]y,p[ \, + \, ]p,y[) - \langle p,y\rangle \, ]p,p[).$$

The choice of b = y yields the generalized form of the Davidon-Fletcher-Powell update (DFP update):

$$\overline{B} = \frac{\rho}{\langle p,y\rangle^2} \, (\langle p,y\rangle^2 B + \langle p,Bp\rangle \, ]y,y[ \, - \langle p,y\rangle \, (]Bp,y[ \, + \, ]y,Bp[))$$

$$+ \frac{1}{\langle p,y\rangle} \, ]y,y[ \, . \tag{3.23}$$

## 3.6 POSITIVE DEFINITE UPDATES

In order to ensure that the direction $v_i = -B_i \nabla F(u_i)$ is a descent direction, we require $B_i$ to be positive definite. This means, that we are looking for update formulas such that positive definiteness is maintained. A basic tool will be the following lemma on eigenvalues of the correction operator:

LEMMA 3.8: Let $T \in L(H)$ be given by

$$T = ]a,b[ + ]c,d[ \quad , \qquad a,b,c,d \in H ,$$

where $\dim TH = 2$.

Then $\lambda = 0$ is an eigenvalue of T with eigenvectors in

$$\{x \in H \mid \langle b,x\rangle a + \langle d,x\rangle c = \Theta\} .$$

Furthermore the solutions

$$\lambda_{+/-} = \Lambda \pm (\Delta + \Lambda^2)^{1/2} \quad \text{of} \quad \lambda^2 = 2\Lambda\lambda + \Delta \tag{3.25}$$

with

$$2\Lambda = \langle a,b\rangle + \langle c,d\rangle , \tag{3.26}$$

$$\Delta = \langle a,d\rangle \langle b,c\rangle - \langle a,b\rangle \langle c,d\rangle \tag{3.27}$$

are also eigenvalues .

Proof: $\lambda$ is an eigenvalue of T if and only if there exists an $x \in H$ such that $x \neq \Theta$ and

$$Tx = \langle b,x\rangle a + \langle d,x\rangle c = \lambda x . \tag{3.28}$$

The statement for the eigenvalue 0 is evident. Otherwise (3.28) implies the existence of $\zeta \in \mathbb{R}$ such that

$$x = a + \zeta c \quad \text{or} \quad x = \zeta a + c .$$

For definiteness, suppose the first equality is true. Replacing x in (3.28) by $a + \zeta c$ and using the linear independence of a and c, we obtain

$$\langle a,b\rangle + \zeta \langle b,c\rangle = \lambda \qquad \text{and} \qquad \langle a,d\rangle + \zeta \langle c,d\rangle = \lambda\zeta , \tag{3.29}$$

After some calculations (3.29) yields the quadratic expression (3.25) for $\lambda$.

DEFINITION 3.9: A linear operator $S \in L(H)$ is called strictly positive definite if and only if there is $\alpha > 0$ such that

$$\langle u, Su\rangle \geq \alpha \langle u,u\rangle \qquad \text{for all} \quad u \in H . \tag{3.30}$$

THEOREM 3.10: Let $B \in L(H)$ be self-adjoint and strictly positive definite, $p,z \in H$, $p,z \neq \Theta$.

Then $\bar{B} = \rho B + T$, $T \in R(p,z)$, is strictly positive definite if and only if with $T$ given by (3.17) $\rho > 0$,

$$(\langle b,B^{-1}y\rangle^2 + (\rho\zeta - \langle b,B^{-1}b\rangle)\langle z,B^{-1}y\rangle)(\zeta\langle z,p\rangle + \langle b,p\rangle^2)^{-1} > 0 \tag{3.31}$$

and

$$\rho + \frac{1}{2}(\zeta \langle z,p\rangle + \langle b,p\rangle^2)^{-1} \quad (-\langle z,p\rangle \langle b,B^{-1}b\rangle + \zeta \langle z,B^{-1}z\rangle$$
$$+ 2 \langle b,p\rangle \langle z,B^{-1}b\rangle ) \geq 0 . \tag{3.32}$$

Proof: $\bar{B}$ is strictly positive definite if and only if there is $\alpha_1 > 0$ such that

$$\langle u, \bar{B}u\rangle = \rho \langle u, Bu\rangle + \zeta_1 (\zeta \langle z,u\rangle^2 - \langle z,p\rangle \langle b,u\rangle^2$$
$$+ 2 \langle b,p\rangle \langle b,u\rangle \langle z,u\rangle) \geq \alpha_1 \langle u,Bu\rangle \tag{3.33}$$

for all $u \in H$ where

$$\zeta_1 = (\zeta \langle z,p\rangle + \langle b,p\rangle^2)^{-1} . \tag{3.34}$$

Setting $\bar{z} = B^{-1/2} z$, $\bar{b} = B^{-1/2} b$, $\bar{p} = B^{1/2} p$ (3.33) is equivalent to

$$\zeta_1 (\zeta \langle \bar{z},v\rangle^2 - \langle z,p\rangle \langle \bar{b},v\rangle^2 + 2 \langle b,p\rangle \langle \bar{b},v\rangle \langle \bar{z},v\rangle)$$
$$\geq (\alpha_1 - \rho) \langle v,v\rangle \tag{3.35}$$

for all $v \in H$.

Hence, we determine the minimum of $\varphi : H \to \mathbb{R}$ on the unit sphere $\{v \in H \mid \langle v,v\rangle = 1\}$ where $\varphi(v)$ equals the left side of the inequality in (3.35). Since the minimum of $\varphi$ on the unit ball exists due to weak compactness and weak continuity arguments, and because it is equal to the minimum on the unit sphere, we have by the Lagrange multiplier rule for each optimal $\hat{v}$ with $\langle \hat{v},\hat{v}\rangle = 1$ the existence of a real $\mu \in \mathbb{R}$ such that

$$\zeta_1 (\zeta \langle \bar{z},\hat{v}\rangle \bar{z} - \langle z,p\rangle \langle \bar{b},\hat{v}\rangle \bar{b}$$
$$+ \langle b,p\rangle (\langle \bar{b},\hat{v}\rangle \bar{z} + \langle \bar{z},\hat{v}\rangle \bar{b})) = \mu\hat{v} . \tag{3.36}$$

Multiplying (3.36) by $\hat{v}$, we realize that $\mu$ is the minimal value of the problem. On the other hand, obviously (3.36) is an eigenvalue problem of type (3.28). We wish to determine conditions such that the eigenvalue $\mu$ is greater than $-\rho$. By Lemma 3.8 $\mu = 0$ is an eigenvalue, hence $\rho > 0$.

For $\mu \neq 0$ let us compute $\Lambda$ and $\Delta$ given in (3.26) and (3.27).

$$\Lambda = \frac{1}{2} \zeta_1 \, (\zeta \, \langle\bar{z},\bar{z}\rangle + 2 \langle b,p\rangle \langle\bar{b},\bar{z}\rangle - \langle z,p\rangle \langle\bar{b},\bar{b}\rangle), \qquad (3.37)$$

$$\Delta = \zeta_1^{\,2} \, [(- \langle z,p\rangle \langle\bar{b},\bar{z}\rangle + \langle b,p\rangle \langle\bar{z},\bar{z}\rangle)(\zeta\langle\bar{b},\bar{z}\rangle + \langle b,p\rangle\langle\bar{b},\bar{b}\rangle)$$

$$- (\zeta\langle\bar{z},\bar{z}\rangle + \langle b,p\rangle \langle\bar{b},\bar{z}\rangle)(\langle b,p\rangle \langle\bar{b},\bar{z}\rangle - \langle\bar{b},\bar{b}\rangle \langle z,p\rangle)]$$

$$= \zeta_1 \, (\langle\bar{z},\bar{z}\rangle \langle\bar{b},\bar{b}\rangle - \langle\bar{b},\bar{z}\rangle^2) \, . \qquad (3.38)$$

First, we check $\Delta + \Lambda^2 > 0$. This holds if $\langle z,p\rangle = 0$. If $|\langle z,p\rangle|$ is greater than zero, then a short computation shows $\Delta + \Lambda^2 > 0$ . Define

$$r = \frac{1}{\sqrt{|\langle z,p\rangle|}} \, ( \langle b,p\rangle \, \bar{z} - \langle z,p\rangle \, \bar{b} ) \, .$$

Then

$$\langle r,r\rangle \langle\bar{z},\bar{z}\rangle - \langle r,\bar{z}\rangle^2 = |\langle z,p\rangle| (\langle\bar{z},\bar{z}\rangle \langle\bar{b},\bar{b}\rangle - \langle\bar{b},\bar{z}\rangle^2)$$

and

$$\Delta + \Lambda^2 = \zeta_1 |\langle z,p\rangle|^{-1} (\langle r,r\rangle \langle\bar{z},\bar{z}\rangle - \langle r,\bar{z}\rangle^2)$$

$$+ \frac{1}{4} \zeta_1^{\,2} \, ((\zeta_1 \langle z,p\rangle)^{-1} \langle\bar{z},\bar{z}\rangle - \operatorname{sgn} \langle z,p\rangle \langle r,r\rangle)^2$$

$$= - (\zeta_1 |\langle z,p\rangle|^{-1} \langle r,\bar{z}\rangle^2$$

$$+ \frac{1}{4} \zeta_1^{\,2} \, ((\zeta_1 \langle z,p\rangle)^{-1} \langle\bar{z},\bar{z}\rangle + \operatorname{sgn} \langle z,p\rangle \langle r,r\rangle)^2$$

which is nonnegative, independent of the sign of $\zeta_1$.

Hence by Lemma 3.8 the eigenvalues of (3.36) are greater than $-\rho$ if and only if

$$\rho + \Lambda \geq (\Delta + \Lambda^2)^{1/2} \, .$$

This is equivalent to $\rho + \Lambda \geq 0$, i.e. (3.32), and

$$\rho^2 + 2\rho\Lambda - \Delta \geq 0 \, .$$

$$\rho^2 + 2\rho\Lambda - \Delta = \zeta_1 \; (\rho^2 \zeta_1^{-1} + \rho(\zeta <\bar{z},\bar{z}> + 2 <b,p> <\bar{b},\bar{z}>$$

$$- <z,p> <\bar{b},\bar{b}>) + <\bar{b},\bar{z}>^2 - <\bar{z},\bar{z}> <\bar{b},\bar{b}>)$$

$$= \zeta_1 \; ((\rho <b,p> + <\bar{b},\bar{z}>)^2 + (\rho\zeta - <\bar{b},\bar{b}>)(\rho <z,p> + <\bar{z},\bar{z}>))$$

and with $\bar{y} = B^{-1/2} y = B^{-1/2} z + \rho B^{1/2} p = \bar{z} + \rho\bar{p}$

$$= (<\bar{b},\bar{y}>^2 + (\rho\zeta - <\bar{b},\bar{b}>) <\bar{z},\bar{y}>)(\zeta <\bar{z},\bar{p}> + <\bar{b},\bar{p}>^2)^{-1}$$

which gives condition (3.31).

In practice, condition (3.32) is often fulfilled if (3.31) holds.

<u>COROLLARY 3.11:</u> If in Theorem 3.10

$$\zeta_1^{-1} = \zeta <z,p> + <b,p>^2 > 0 \; ,$$

then (3.32) is redundant and (3.31) reduces to

$$<b, B^{-1}y>^2 + (\rho\zeta - <b, B^{-1}b>) <z, B^{-1}y> > 0 \; . \qquad (3.39)$$

If $\zeta_1$ is positive, then with the notation in the proof,

$$\rho \; (\rho + 2\Lambda) \geq \Delta \geq 0 \; ,$$

and since $\rho > 0$

$$\rho + 2\Lambda \geq 0$$

and also

$$\rho + \Lambda \geq 0 \; ,$$

i.e. (3.32) is satisfied.

If we choose $\zeta = 0$, then $\zeta_1 = <b,p>^{-2} > 0$ for all $b \in H$ with $<b,p> \neq 0$ and, by Corollary 3.11, it remains to check (3.39):

$$<b, B^{-1}y>^2 \geq <b, B^{-1}b> <z, B^{-1}y>$$

$$= <b, B^{-1}b> (<y, B^{-1}y> - \rho <p,y>) \; .$$

It is known that the unscaled version ($\rho = 1$) of the PSB update (3.22) for b = p does not necessarily maintain the positive definiteness of the update operators $B_i$. This however is true if the parameter $\rho$ is chosen large enough. For the general class of updates considered in (3.22) we have

PROPOSITION 3.12: Let the update formula be given by (3.22). Then positive definiteness is maintained if and only if

$$\rho \langle p,y\rangle \langle b, B^{-1}b\rangle \geq \langle b, B^{-1}b\rangle \langle y, B^{-1}y\rangle - \langle b, B^{-1}y\rangle^2 . \quad (3.40)$$

Since $\langle p,y\rangle > 0$ (see Section 1.5), (3.40) gives a lower bound for the choice of $\rho$. Obviously, (3.40) is satisfied, if $b = y$ and $\rho > 0$, the scaled DFP update. This selection gives the largest interval for $\rho$.

Another way of fulfilling

$$\zeta_1^{-1} = \zeta \langle y,p\rangle - \zeta\rho \langle Bp,p\rangle + \langle b,p\rangle^2 > 0$$

is to choose

$$\zeta = \frac{\langle b,p\rangle^2}{\rho \langle p,Bp\rangle} , \quad \rho > 0 . \quad (3.41)$$

Then (3.39) is

$$\langle b, B^{-1}y\rangle^2 + (\langle b,p\rangle^2 \langle p,Bp\rangle^{-1} - \langle b, B^{-1}b\rangle) \langle z, B^{-1}y\rangle > 0 .$$

This is satisfied for $b = Bp$. With (3.41) it gives the scaled version of the Broyden-Fletcher-Goldfarb-Shanno update (BFGS update)

$$\overline{B} = \rho \left(B - \frac{1}{\langle p,Bp\rangle} ]Bp, Bp[\right) + \frac{1}{\langle p,y\rangle} ]y,y[ .$$

## 3.7 COMMENTS

Section 3.1 - 3.6: Variable metric methods were presented first by Davidon [1] and Fletcher and Powell [2]. For an article which reviews this field extensively and gives a detailed list of references we refer the reader to [3].

Various classes of updates have been introduced, see for example [4] - [7]. Scaling variable metric methods were proposed in [8] and [9].

The BFGS update is known to be one of the most numerically efficient procedures. It was given independently by Broyden [10], Fletcher [5], Goldfarb [11] and Shanno [12].

# 4 Global convergence of variable metric methods

## 4.1 ZOUTENDIJK'S GENERAL THEOREM

A general convergence condition has been given by Zoutendijk. We shall use it in connection with the step length rules of Chapter 2. Let us impose assumption $A_1$ of Section 2.1. We define cond B as the root of the condition number for a self-adjoint operator B.

DEFINITION 4.1: Let $B \in L(H)$ be such that there exist $\beta_-, \beta_+ \in \mathbb{R}$ with $\beta_+ \geq \beta_- > 0$, $\beta_+$ minimal, $\beta_-$ maximal,

$$\beta_- \langle u,u\rangle \leq \langle u,Bu\rangle \leq \beta_+ \langle u,u\rangle \tag{4.1}$$

for all $u \in H$. Then the number cond B is defined by

$$\text{cond } B = \frac{\beta_+}{\beta_-} .$$

The preference for matrices will small condition number is well known in numerical analysis. For convergence analysis of variable metric methods, estimates of the condition number of the update operators are also very useful.

We prove a general convergence theorem due to Zoutendijk.

THEOREM 4.2: Let $A_1$ be fulfilled with E = H, a Hilbert space. Assume that $\{u_i\}_{\mathbb{N}}$ is a sequence of iterates constructed by a variable metric method as presented in Section 3.1 with descent directions $\{v_i\}_{\mathbb{N}}$,

$$v_i = B_i^{-1} \nabla F(u_i) , \tag{4.2}$$

$B_i$ self-adjoint operators in L(H). If the condition numbers of $B_i$ are uniformly bounded,

$$\text{cond } B_i \leq \mu \qquad \text{for all } i \in \mathbb{N} \tag{4.3}$$

and if the Zoutendijk condition

$$\sum_{i=0}^{\infty} \langle \nabla F(u_i),\, v_i \rangle^2 \langle v_i, v_i \rangle^{-1} < \infty \tag{4.4}$$

holds, then

$$\lim_{i \to \infty} \| \nabla F(u_i) \| = O \;. \tag{4.5}$$

<u>Proof:</u> Insert (4.2) into (4.4)

$$\infty > \sum_{i=0}^{\infty} \langle \nabla F(u_i),\, B_i^{-1} \nabla F(u_i) \rangle^2 \langle B_i^{-1} \nabla F(u_i),\, B_i^{-1} \nabla F(u_i) \rangle^{-1}$$

$$\geq \sum_{i=0}^{\infty} \langle \nabla F(u_i),\, B_i^{-1} \nabla F(u_i) \rangle \, \beta_{i}$$

$$\geq \sum_{i=0}^{\infty} \| \nabla F(u_i) \|^2 \; (\text{cond } B_i)^{-1} \;.$$

From this we obtain with (4.3) condition (4.5).

The Zoutendijk condition has been discussed in Chapter 2 and can be satisfied by several step length rules. The main purpose of this chapter will be to provide estimates on the condition number of variable metric updates.

## 4.2 ESTIMATES OF THE CONDITION NUMBER

We have seen that bounds on the condition numbers of B play an essential role in the convergence proof. In the next theorem we give an estimate for a general class R(p,z) of self-adjoint updates. Hence the correction term T is given by $\zeta \in \mathbb{R}$, $b \in H$, $\zeta_1^{-1} = \zeta \langle z,p \rangle + \langle b,p \rangle^2 \neq O$ and

$$T = \zeta_1 \left( \zeta \, ]z,z[ \; - \langle z,p \rangle \, ]b,b[ \; + \langle b,p \rangle \, (]b,z[ \; + \; ]z,b[) \right). \tag{4.6}$$

<u>LEMMA 4.3:</u> Let $B \in L(H)$, positive definite, be given with finite condition number. Then with $T \in R(p,z)$ represented as in (4.6), we have for positive definite operators $\overline{B} = \rho B + T$ the following estimates:

If $\zeta_1 \langle z,p \rangle \geq O$ , then

$$\operatorname{cond} \bar{B} \le \operatorname{cond} B \frac{\rho + \Lambda + (\Delta + \Lambda^2)^{1/2}}{\rho + \Lambda - (\Delta + \Lambda^2)^{1/2}} \quad . \tag{4.7}$$

If $\langle z,p\rangle,\ -\zeta_1 > 0$, then

$$\operatorname{cond} \bar{B} \le \operatorname{cond} B \ \frac{1}{\rho} \ (\rho + \Lambda + (\Delta + \Lambda^2)^{1/2}) . \tag{4.8}$$

If $\langle z,p\rangle,\ -\zeta_1 < 0$, then

$$\operatorname{cond} \bar{B} \le \operatorname{cond} B \, \rho \ (\rho + \Lambda + (\Delta + \Lambda^2)^{1/2})^{-1} \ . \tag{4.9}$$

<u>Proof:</u> Similarly as in the proof of Theorem 3.10 we minimize or maximize $\langle u, \bar{B}u\rangle$ on the sphere $S_B = \{u \in H \mid \langle u, Bu\rangle = 1\}$ and following the same proof, we find with Lemma 3.8 that there are three possible values for the minimum and maximum:

$$\rho, \quad \rho + \Lambda + (\Delta + \Lambda^2)^{1/2}, \quad \rho + \Lambda - (\Delta + \Lambda^2)^{1/2},$$

where $\Lambda$, $\Delta$ are given by (3.37), (3.38). Since

$$\sup_{\|u\|=1} \langle u, \bar{B}u\rangle \le \sup_{\|u\|=1} \langle u,Bu\rangle \ \sup_{v\in H} \langle v,\bar{B}v\rangle \langle v,Bv\rangle^{-1}$$

and a similar estimate holds for the infima,

$$\operatorname{cond} \bar{B} \le \operatorname{cond} B \ \sup_{v \in S_B} \langle v, \bar{B}v\rangle \ (\inf_{v \in S_B} \langle v, \bar{B}v\rangle)^{-1}$$

$$= \operatorname{cond} B \ \frac{\max\{\rho,\ \rho + \Lambda + \eta\}}{\min\{\rho,\ \rho + \Lambda - \eta\}}, \quad \eta = (\Delta + \Lambda^2)^{1/2} \quad . \tag{4.10}$$

If $\langle z,p\rangle = 0$, then $\zeta_1 > 0$ and $\Delta > 0$ . Hence $\Lambda + \eta \ge 0$ and $\Lambda - \eta \le 0$, and (4.7) holds.

Assuming $\langle z,p\rangle \ne 0$ we define r by

$$r = ( |\langle z,p\rangle| )^{-1/2} ( \langle b,p\rangle \bar{z} - \langle z,p\rangle \bar{b} ).$$

Then

$$\Lambda = \frac{1}{2} (\langle z,p\rangle^{-1} \langle \bar{z},\bar{z}\rangle - \operatorname{sgn}(\langle z,p\rangle) \zeta_1 \langle r,r\rangle ) ,$$

$$\Delta = \zeta_1 \ |\langle z,p\rangle|^{-1} \ (\langle r,r\rangle \langle \bar{z},\bar{z}\rangle - \langle r,\bar{z}\rangle^2 ) .$$

Consider $<z,p> > 0$. If $\zeta_1 > 0$, then $\Delta \geq 0$ and as before we obtain (4.7). Negative $\zeta_1$ implies $\Delta \leq 0$ and $\Lambda \pm \eta \geq 0$.

Therefore (4.8) is true. Similarly, the case $<z,p> < 0$ can be discussed. This completes the proof.

If F is a quadratic functional,

$$F(u) = \frac{1}{2} <u,Qu> + <c,u> + d \ , \quad Q \in L(H),\ c,d \in H \ , \tag{4.11}$$

Q self-adjoint and strictly positive definite, then

$$y_i = \nabla F(u_{i+1}) - \nabla F(u_i) = Q\,(u_{i+1} - u_i) = Qp_i \ . \tag{4.12}$$

This relation will be used to derive another estimate of the condition number.

<u>LEMMA 4.4:</u> Suppose F is a quadratic of the form (4.11). Let $B \in L(H)$ be self-adjoint and with $\eta_-, \eta_+ \in \mathbb{R}$, $\eta_- > 0$, such that

$$\eta_- <u, Qu> \leq <u, Bu> \leq \eta_+ <u, Qu> \quad \text{for all } u \in H \ .$$

Then with $T \in R(p,z)$ given by (4.6) and

$$b = \gamma y + \pi Bp \in H \ , \qquad \gamma, \pi \in \mathbb{R} \ , \tag{4.13}$$

such that

$$\rho^2\,(\zeta + <p,y>\,\gamma^2) = <p,y>\,(\pi + \gamma\rho)^2 + \rho <p, Bp>\,\pi^2 \ , \tag{4.14}$$

for $\overline{B} = \rho B + T$ there are $\overline{\eta}_+, \overline{\eta}_- > 0$ fulfilling

$$\overline{\eta}_- <u, Qu> \leq <u, \overline{B}u> \leq \overline{\eta}_+ <u, Qu> \quad \forall u \in H \tag{4.15}$$

and

$$\overline{\eta}_-, \overline{\eta}_+ \in [\rho\eta_-, \rho\eta_+] \cup \{1\} \ . \tag{4.16}$$

<u>Proof:</u> We minimize $<u, \overline{B}u>$ on the sphere $S_Q = \{u \in H \mid <u, Qu> = 1\}$. By the multiplier rule there exists for each optimal $\hat{u} \in S_Q$ a real $\mu$ such that

$$\rho B\hat{u} + T\hat{u} = \mu Q\hat{u} \ .$$

Multiplying this equation by p and using the quasi-Newton equation $Tp = z = y - \rho Bp$ we obtain

$$<y,\hat{u}> = \mu <p, Q\hat{u}> \ .$$

If $<y,\hat{u}> \neq 0$, we deduce $\mu = 1$ from (4.12).

On the other hand suppose $0 = \langle y,\hat{u}\rangle = \langle z,\hat{u}\rangle - \rho \langle p, Bp\rangle$. Then

$$\zeta_1^{-1}\langle \hat{u}, T\hat{u}\rangle = \zeta \langle z,\hat{u}\rangle^2 - \langle z,p\rangle \langle b,\hat{u}\rangle^2 + 2 \langle b,p\rangle \langle b,\hat{u}\rangle \langle z,\hat{u}\rangle$$

$$= \zeta\rho^2 \langle Bp, \hat{u}\rangle^2 - \langle z,p\rangle \langle b,\hat{u}\rangle^2 - 2\rho \langle b,p\rangle \langle b,\hat{u}\rangle \langle Bp, \hat{u}\rangle$$

with (4.13) , $\langle b,\hat{u}\rangle = \pi \langle Bp, \hat{u}\rangle$ ,

$$= (\zeta\rho^2 - \pi^2 \langle z,p\rangle - 2\rho\pi \langle b,p\rangle) \langle Bp, \hat{u}\rangle^2$$

$$= (\zeta\rho^2 - \pi^2 \langle p,y\rangle - 2\rho\pi\gamma\langle p,y\rangle - \rho\pi^2 \langle p, Bp\rangle) \langle Bp, \hat{u}\rangle^2$$

$$= 0 \qquad \text{by (4.14).}$$

Therefore, for $\hat{u}$ with $\langle y,\hat{u}\rangle = 0$,

$$\langle \hat{u}, \bar{B}\hat{u}\rangle = \rho \langle \hat{u}, B\hat{u}\rangle$$

and

$$\mu \in \rho\, [\eta_-, \eta_+]$$

## 4.3 BFGS METHOD

In the finite dimensional case the special structure of the BFGS method allows a global convergence proof. We shall extend this proof to the infinite dimensional case with step length rules of Chapter 2.

THEOREM 4.5: Let $(A_2)$ be fulfilled. Then the sequence generated by the BFGS method converges globally to the optimum if the step length rule satisfies the Zoutendijk condition.

In the BFGS method, the update is given by

$$B_{i+1} = B_i + \frac{]y_i, y_i[}{\langle p_i, y_i\rangle} - \frac{]B_ip_i, B_ip_i[}{\langle p_i, B_ip_i\rangle} \tag{4.17}$$

and with (3.5)

$$B_{i+1} = B_i \left(I + \frac{]B_i^{-1} y_i, y_i[}{\langle p_i, y_i\rangle} - \frac{]p_i, B_ip_i[}{\langle p_i, B_ip_i\rangle}\right) \tag{4.18}$$

or by replacing $B_i$ by the same formula with a smaller index using (4.17)

$$B_o^{-1} B_{i+1} = I + \sum_{j=0}^{i} \left( \frac{]B_o^{-1} y_j, y_j[}{\langle p_j, y_j \rangle} - \frac{]B_o^{-1} B_j p_j, B_j p_j[}{\langle p_j, B_j p_j \rangle} \right) \tag{4.19}$$

and using (4.18)

$$B_o^{-1} B_{i+1} = \prod_{j=0}^{i} \left( I + \frac{]B_j^{-1} y_j, y_j[}{\langle p_j, y_j \rangle} - \frac{]p_j, B_j p_j[}{\langle p_j, B_j p_j \rangle} \right). \tag{4.20}$$

The formula for the inverse $H_{i+1}$ of $B_{i+1}$ is (Theorem 3.3)

$$H_{i+1} = H_i + \langle p_i, y_i \rangle^{-1} (]p_i - H_i y_i, p_i[ + ]p_i, p_i - H_i y_i[)$$
$$- \langle p_i - H_i y_i, y_i \rangle \langle p_i, y_i \rangle^{-2} ]p_i, p_i[ . \tag{4.21}$$

<u>LEMMA 4.6:</u> The operators $B_o^{-1} B_k$, $k = 0,\dots,i+1$, $B_k$ given by (4.17) result in

$$T_i = \text{span}\,\{B_o^{-1} y_j, B_o^{-1} B_j p_j \in H \mid j = 0,\dots,i\}$$

and

$$S_i = \text{span}\,\{B_j^{-1} y_j, p_j \in H \mid j = 0,\dots,i\}$$

invariant. Furthermore

$$T_i = S_i$$

for all $i \in \mathbb{N}$.

<u>Proof:</u> (4.19) and (4.20) imply that

$$B_o^{-1} B_k (T_i) \subseteq T_i , \quad B_o^{-1} B_k (S_i) \subseteq S_i$$

for all $k = 0,\dots,i+1$.

$T_o = S_o$ is true by definition.

Let us assume $T_i = S_i$ for some $i \geq 0$. Then we have to show that

$$T_{i+1} = \text{span}\,\{B_o^{-1} y_{i+1}, B_o^{-1} B_{i+1} p_{i+1}, T_i\}$$
$$= \text{span}\,\{B_{i+1}^{-1} y_{i+1}, p_{i+1}, S_i\} = S_{i+1} .$$

By (4.19)

$$B_o^{-1} B_{i+1} p_{i+1} - p_{i+1} \in T_i = S_i ,$$

hence $B_o^{-1} B_{i+1} p_{i+1} \in S_{i+1}$ and $p_{i+1} \in T_{i+1}$. Successive application of (4.21) gives

$$B_{i+1}^{-1} y_{i+1} = B_i^{-1} y_{i+1} + s_o = B_{i-1}^{-1} y_{i+1} + s_1 = \dots$$
$$= B_o^{-1} y_{i+1} + s_i ,$$

for some $s_k \in S_i = T_i$, $k = 0,\dots,i$. Therefore

$$B_{i+1}^{-1} y_{i+1} \in T_{i+1} \quad \text{and} \quad B_o^{-1} y_{i+1} \in S_{i+1} ,$$

which completes the proof.

Proof of Theorem 4.5 :

Let $i \in \mathbb{N}$ be fixed. By Lemma 4.6 we can restrict the operators $B_o^{-1} B_k$, $k = 0,\dots,i+1$, to the finite dimensional subspace $T_i$. We denote the matrix which is obtained by the representation of the restricted operator $B_o^{-1} B_k$ by $M(B_o^{-1} B_k)$. $\lambda$ is eigenvalue of $B_o^{-1} B_k$ with eigenvector in $T_i$ if and only if $\lambda$ is eigenvalue of $B_o^{-1/2} B_k B_o^{-1/2}$ with eigenvector in $B_o^{1/2} T_i$. Since $B_o^{-1/2} B_k B_o^{-1/2}$ is positive definite on $B_o^{1/2} T_i$ there exists a full set of positive eigenvalues. The same argument applies to $M(]B_j^{-1} y_j, y_j[)$ and $M(]p_j, B_j p_j[)$.

The additivity of the traces yields with (4.19)

$$\operatorname{tr} M(B_o^{-1} B_{i+1}) = \operatorname{tr} M(I) + \sum_{j=0}^{i} (\langle p_j, y_j\rangle^{-1} \operatorname{tr} M(]B_o^{-1} y_j, y_j[)$$
$$- \langle p_j, B_j p_j\rangle^{-1} \operatorname{tr} M(]B_o^{-1} B_j p_j, B_j p_j[)). \tag{4.22}$$

Rank-one operators $]t,a[$, $t \in T_i$, $a \in H$, have only one (possibly) nonzero eigenvalue $\langle t,a\rangle$. Hence (4.22) implies

$$\operatorname{tr} M(B_o^{-1} B_{i+1}) = \operatorname{tr} M(I)$$
$$+ \sum_{j=0}^{i} \left( \frac{\langle y_j, B_o^{-1} y_j\rangle}{\langle p_j, y_j\rangle} - \frac{\langle B_j p_j, B_o^{-1} B_j p_j\rangle}{\langle p_j, B_j p_j\rangle} \right) . \tag{4.23}$$

Using $\operatorname{tr} M(I) \le 2(i + 1)$ and $0 < \operatorname{tr} M(B_o^{-1} B_{i+1})$ because of the positivity of the eigenvalues of $M(B_o^{-1} B_{i+1})$ we obtain from (4.23)

$$\sum_{j=0}^{i} \frac{\langle B_j p_j, B_o^{-1} B_j p_j \rangle}{\langle p_j, B_j p_j \rangle} \le \sum_{j=0}^{i} \frac{\langle y_j, B_o^{-1} y_j \rangle}{\langle p_j, y_j \rangle} + 2(i+1) .$$

Since $B_o$ has a finite condition number and $\langle p,y \rangle \le c_1 \langle y,y \rangle$, cf. Section 1.5, for some $c_2 > 0$

$$\sum_{j=0}^{i} \frac{\langle B_j p_j, B_j p_j \rangle}{\langle p_j, B_j p_j \rangle} \le c_2 (i+1) . \tag{4.24}$$

This formula, being derived from (4.19) and the traces of the operators restricted to $T_i$, will be used later.

At this point we consider (4.20) and the product formula for determinants

$$\det M (B_o^{-1} B_{i+1}) = \prod_{j=0}^{i} \det M \left(I + \frac{]B_j^{-1} y_j, y_j[}{\langle p_j, y_j \rangle} - \frac{]p_j, B_j p_j[}{\langle p_j, B_j p_j \rangle}\right) . \tag{4.25}$$

Omitting indices, we derive from Lemma 3.8 that each of the matrices appearing in the product in (4.25) has eigenvalues 1 and

$$1 + \frac{1}{2} \left( \frac{\langle y, B^{-1} y \rangle}{\langle p,y \rangle} - 1 \right) \pm \left( - \frac{\langle p,y \rangle}{\langle p, Bp \rangle} + \frac{1}{4} \left( \frac{\langle y, B^{-1} y \rangle}{\langle p,y \rangle} + 1 \right)^2 \right)^{1/2} .$$

Therefore, the determinant is given by

$$\det M \left(I + \frac{]B^{-1} y, y[}{\langle p,y \rangle} - \frac{]p, Bp[}{\langle p, Bp \rangle}\right)$$

$$= \left(1 + \frac{1}{2} \left( \frac{\langle y, B^{-1} y \rangle}{\langle p,y \rangle} - 1 \right)\right)^2 + \frac{\langle p,y \rangle}{\langle p, Bp \rangle} - \frac{1}{4} \left( \frac{\langle y, B^{-1} y \rangle}{\langle p,y \rangle} + 1 \right)^2$$

$$= \langle p,y \rangle \langle p, Bp \rangle^{-1} .$$

Hence from (4.25) and using the inequality of the geometric-arithmetic mean on the eigenvalues of $M(B_o^{-1} B_{i+1})$

$$\prod_{j=0}^{i} \frac{\langle y_j, p_j \rangle}{\langle p_j, B_j p_j \rangle} = \det M(B_o^{-1} B_{i+1}) \le \left( \frac{\operatorname{tr} M(B_o^{-1} B_{i+1})}{i+1} \right)^{i+1} .$$

With (4.23) this yields for some constant $c_3 > 0$

$$\prod_{j=0}^{i} \frac{\langle y_i, p_i \rangle}{\langle p_j, B_j p_j \rangle} \leq c_3^{i+1} . \tag{4.26}$$

If we apply the inequality of the geometric-arithmetic mean to (4.24), it follows

$$\prod_{j=0}^{i} \frac{\langle B_j p_j, B_j p_j \rangle}{\langle p_j, B_j p_j \rangle} \leq \left( \frac{1}{i+1} \sum_{j=0}^{i} \frac{\langle B_j p_j, B_j p_j \rangle}{\langle p_j, B_j p_j \rangle} \right)^{i+1}$$

$$\leq c_2^{i+1} , \tag{4.27}$$

hence from (4.26) and (4.27) with $c_4 = \frac{1}{c_2 c_3}$

$$\prod_{j=0}^{i} \eta_j \geq c_4^{i+1} \tag{4.28}$$

where

$$\eta_j = \frac{\langle p_j, B_j p_j \rangle^2}{\langle y_j, p_j \rangle \langle B_j p_j, B_j p_j \rangle} . \tag{4.29}$$

After these preparations we come to the convergence proof itself.

The Zoutendijk condition (4.4) yields

$$\sum_{i=0}^{\infty} \frac{\langle \nabla F(u_i), B_i^{-1} \nabla F(u_i) \rangle^2}{\langle B_i^{-1} \nabla F(u_i), B_i^{-1} \nabla F(u_i) \rangle} < \infty$$

and since $p_i = \alpha_i B_i^{-1} \nabla F(u_i)$, we have

$$\sum_{i=0}^{\infty} \alpha_i^2 \frac{\langle B_i p_i, p_i \rangle^2}{\langle p_i, p_i \rangle} < \infty . \tag{4.30}$$

With $\eta_i$ defined by (4.29) and $\langle y_i, p_i \rangle \geq \gamma \langle p_i, p_i \rangle$ (see Section 1.5) we obtain from (4.30)

$$\sum_{i=0}^{\infty} \eta_i \, \| \nabla F(u_i) \|^2 < \infty , \tag{4.31}$$

If $\alpha \in \mathbb{R}$, then $1 + \alpha \leq \exp \alpha$ and

$$\prod_{i=0}^{n} (1 + \alpha_i) \leq \exp \left( \sum_{i=0}^{n} \alpha_i \right) , \quad \alpha_i \geq 0, \ i = 0, \ldots, n .$$

Since the sum in (4.31) is finite,

$$\lim_{n\to\infty} \prod_{i=0}^{n} (1 + \|\nabla F(u_i)\|^2 \eta_i)$$

$$\le \exp \left( \sum_{i=0}^{\infty} \|\nabla F(u_i)\|^2 \eta_i \right) < \infty . \tag{4.32}$$

For each $n \in \mathbb{N}$

$$\prod_{i=0}^{n} (1 + \|\nabla F(u_i)\|^2 \eta_i)$$

$$= \prod_{i=0}^{n} \eta_i \prod_{i=0}^{n} \|\nabla F(u_i)\|^2 \ (1 + \|\nabla F(u_i)\|^{-2} \eta_i^{-1})$$

and from (4.28)

$$\ge c_4^{n+1} \prod_{i=0}^{n} \|\nabla F(u_i)\|^2 \ (1 + \|\nabla F(u_i)\|^{-2} \eta_i^{-1})$$

and with (4.32)

$$\lim_{n\to\infty} \prod_{i=0}^{n} (c_4 \|\nabla F(u_i)\|^2 \ (1 + \|\nabla F(u_i)\|^{-2} \eta_i^{-1})) < \infty. \tag{4.33}$$

From (4.32) we have that

$$\|\nabla F(u_i)\|^{-2} \eta_i^{-1} \to \infty ,$$

hence (4.33) implies

$$\lim_{i\to\infty} \|\nabla F(u_i)\| = 0 .$$

This completes the proof.

## 4.4 QUADRATIC FUNCTIONALS

In this section let F be given by

$$F(u) = \langle u, Qu\rangle + \langle c,u\rangle + d \tag{4.34}$$

$Q \in L(H)$ self-adjoint and strictly positive definite, $c,d \in H$.

In the following theorem we shall use the results of Section 4.2.

THEOREM 4.7: Let F be quadratic and $\{u_i\}_{\mathbb{N}}$ a sequence of iterates constructed by a variable metric method where the step length rule satisfies Zoutendijk's condition. The sequence of operators $\{B_i\}$ is given by $B_o \in L(H)$, cond $B_o < \infty$,

$$B_{i+1} = \rho_i B_i + T_i, \quad T_i \in R(p_i, z_i), \quad z_i = y_i - \rho_i B_i p_i,$$

where $T_i$ is represented by (4.6) and

$$b_i = \gamma_i y_i + \pi_i B_i p_i$$

and $\gamma_i, \pi_i, \rho_i, \zeta_i$ fulfill (4.14), $\rho_i > 0$. If

$$\inf_{u \in S_Q} \langle u, B_i u\rangle \leqq \rho_i^{-1} \leq \sup_{u \in S_Q} \langle u, B_i u\rangle, \tag{4.35}$$

then the condition numbers of $B_i$ are uniformly bounded,

$$\text{cond } B_i \leq M \quad \text{for some} \quad M > 0 \quad \text{and all} \quad i \in \mathbb{N}.$$

Proof: At each iteration, all the assumptions of Lemma 4.4 are satisfied, hence (4.16) holds. (4.35) gives

$$\rho\eta_- \leq 1 \leq \rho\eta_+$$

and therefore

$$\bar{\eta}_+ \bar{\eta}_-^{-1} \leq \eta_+ \eta_-^{-1}$$

or

$$\frac{\sup_{v \in S_Q} \langle v, B_i v\rangle}{\inf_{v \in S_Q} \langle v, B_i v\rangle} \leq \frac{\sup_{v \in S_Q} \langle v, B_o v\rangle}{\inf_{v \in S_Q} \langle v, B_o v\rangle} \quad \text{for all} \quad i \in \mathbb{N}.$$

This implies that cond $B_i$ is bounded by some $M > 0$ for all $i \in \mathbb{N}$.

We shall show that condition (4.14) yields for choices of b by (4.17) in general one class of updates:

PROPOSITION 4.8: Each correction $T \in R(p,z)$ where b is given by (4.13) and $\rho, \zeta, \pi, \gamma$ are such that $\rho > 0$ and (4.14) holds, yields the generalized DFP update

$$\bar{B} = \rho \left(B + \frac{\langle p, Bp\rangle}{\langle p,y\rangle^2}\ ]y,y[\ - \frac{1}{\langle p,y\rangle} (]Bp, y[ + ]y, Bp[)\right)$$

$$+ \frac{1}{\langle p,y\rangle}\ ]y,y[\ .$$

<u>Proof:</u> T is given by

$$T = \zeta_1 (\zeta\ ]z,z[\ - \langle z,p\rangle\ ]b,b[\ + \langle b,p\rangle\ (]b,z[ + ]z,b[))$$

such that

$$\zeta_1^{-1} = \zeta \langle z,p\rangle + \langle b,p\rangle^2 \neq 0 ,$$

$$z = y - \rho Bp , \quad b = \gamma y + \pi Bp .$$

(4.14) gives

$$\rho^2 \zeta = \langle p,y\rangle \pi^2 + 2\pi\gamma\rho \langle p,y\rangle + \rho \langle p, Bp\rangle \pi^2 . \qquad (4.36)$$

Hence $\zeta_1^{-1}$ is given by

$$\zeta_1^{-1} = \zeta \langle p,y\rangle - \zeta\rho \langle p, Bp\rangle + \gamma^2 \langle p,y\rangle^2$$

$$+ 2\pi\gamma \langle p,y\rangle \langle p, Bp\rangle + \pi^2 \langle p, Bp\rangle^2$$

replacing $\pi^2 \langle p, Bp\rangle^2$ by (4.36)

$$= \frac{\langle p,y\rangle}{\rho^2} (\zeta\rho^2 + \gamma^2\rho^2 \langle p,y\rangle - \pi^2\rho \langle p, Bp\rangle)$$

using (4.36) once more

$$= \frac{\langle p,y\rangle}{\rho^2} (\gamma^2\rho^2 \langle p,y\rangle + \langle p,y\rangle \pi^2 + 2\pi\gamma\rho \langle p,y\rangle)$$

and

$$\zeta_1^{-1} = \left( \frac{\langle p,y\rangle}{\rho} (\rho\gamma + \pi)\right)^2 . \qquad (4.37)$$

We express z and b in terms of y and Bp in T and collecting terms we obtain as coefficients for

$$]Bp, Bp[\ : \zeta_1 (\zeta\rho^2 - \langle z,p\rangle \pi^2 - 2\rho\pi \langle b,p\rangle) = 0 \quad \text{from (4.36)},$$

$$]y,y[\ : \zeta_1 (\zeta + \langle p,y\rangle \gamma^2 + (2\pi\gamma + \rho\gamma^2) \langle p, Bp\rangle)$$

replacing $\zeta$ with (4.36)

$$= \zeta_1 \left( \frac{\langle p,y\rangle}{\rho^2} + \frac{\langle p, Bp\rangle}{\rho} \right) (\pi^2 + 2\pi\gamma\rho + \rho^2\gamma^2)$$

with (4.37)

$$= \frac{1}{<p,y>} + \rho \frac{<p, Bp>}{<p,y>^2} .$$

$$]Bp, y[ + ]y, Bp[ \ : \zeta_1 \ (-\rho\zeta + (\pi^2 - \gamma\rho\pi)<p,Bp> - (\pi\gamma - \rho\gamma^2)<p,y>)$$

replacing $\pi^2 <p, Bp>$ with (4.36)

$$= \zeta_1 \ (- \frac{<p,y>}{\rho}) \ (\pi + \gamma\rho)^2 = - \frac{\rho}{<p,y>} .$$

Finally, we obtain the generalized DFP update.

This shows that the BFGS update does not satisfy (4.14), since in this case any update leads to the DFP version. In the Proposition, the quadratic form of the cost functional was not utilized.

REMARK: Condition (4.35) is easy to satisfy. Consider for example

$$\rho_i = \frac{<w_i, Qw_i>}{<w_i, Bw_i>} , \quad \text{for some } w_i \in H .$$

The most useful choice of w for (4.35) is $w = p_i$. In this case

$$\rho = \frac{<p, Qp>}{<p, Bp>} = \frac{<p,y>}{<p, Bp>} ,$$

and the two scalar products $<p,y>$, $<p, Bp>$ nevertheless must be calculated for the update of B.

This choice of $\rho$ gives the self-scaling version of the DFP update

$$\bar{B} = \frac{<p,y>}{<p, Bp>} B + \frac{2}{<p,y>} ]y,y[ - \frac{1}{<p, Bp>} (]Bp,y[ + ]y,Bp[).$$

## 4.5 COMMENTS

We cite only a few references of the convergence theory in $\mathbb{R}^n$ that are closely related to our Hilbert space results.

Section 4.1: Zoutendijk's condition plays an important role in the convergence theory of [2] and [3].

Section 4.2: In $\mathbb{R}^n$ estimates of the condition number of a class of scaling variable metric updates are treated in [4].

Section 4.3: A proof of global convergence in $\mathbb{R}^n$ for the BFGS method with an inexact step length rule is given in [5] and is extended to a large class of step length rules in [3].

Section 4.4: In the convergence theory for Hilbert space methods quadratic functionals were usually treated, [6] - [9]. Horwitz and Sarachik [6] and Tokumaru, Adachi and Goto [7] prove convergence of the DFP method, which is extended by Turner and Huntley [8], [9] to classes of the Huang family [10]. An application to optimal control problems is discussed by Edge and Powers [11].

# 5 Convergence rates of variable metric methods

## 5.1 DEFINITION OF CONVERGENCE RATES

In Chapter 4 we have discussed the convergence of variable metric methods. This chapter considers the rate of convergence of $\| u_i - \hat{u} \|$ to zero. We confine this treatment to statements on local convergence in which it is assumed that the initial point u is sufficiently close to $\hat{u}$. This requirement can be satisfied when the iteration is globally convergent, and therefore the distance from an iteration point $u_i$ to $\hat{u}$ is so small that local convergence results are valid.

Four definitions of convergence rates will be treated.

DEFINITION 5.1: Let $\{u_i\}_{\mathbb{N}} \subset H$ and $\hat{u} \in H$ such that

$$\lim_{i \to \infty} u_i = \hat{u} .$$

The convergence rate is called linear if there are $\nu \in (0,1)$, $i_o \in \mathbb{N}$ such that

$$\|u_{i+1} - \hat{u}\| \leq \nu \|u_i - \hat{u}\| \quad \text{for all } i \geq i_o . \tag{5.1}$$

The convergence rate is quadratic if there is $\nu \in \mathbb{R}$, $i_o \in \mathbb{N}$ such that

$$\|u_{i+1} - \hat{u}\| \leq \nu \|u_i - \hat{u}\|^2 \quad \text{for all } i \geq i_o . \tag{5.2}$$

Condition (5.1) implies convergence since $\nu$ is smaller than 1. Moreover, (5.2) yields convergence of $\{u_i\}_{\mathbb{N}}$ to $\hat{u}$ if $u_{i_o}$ is chosen such that $\| u_{i_o} - \hat{u} \| < \nu^{-1}$.

If convergence rate results are possible to prove, one usually attempts to establish linear convergence. Quadratic convergence is rather rare and in general can only be obtained for second order methods which require step-by-step information on the second derivatives.

Between these two latter rates lies superlinear convergence.

DEFINITION 5.2: Let $\{u_i\}_{\mathbb{N}} \subset H$, $\hat{u} \in H$ such that

$$\lim_{i \to \infty} u_i = \hat{u} \text{ at a linear rate.}$$

The convergence rate is called superlinear if

$$\lim_{i\to\infty} \frac{\|u_{i+1} - \hat{u}\|}{\|u_i - \hat{u}\|} = 0. \tag{5.3}$$

The convergence rate is weakly superlinear if for each $\ell \in H$

$$\lim_{i \to \infty} \frac{\langle \ell, u_{i+1} - \hat{u}\rangle}{\|u_i - \hat{u}\|} = 0 . \tag{5.4}$$

Both conditions (5.3) and (5.4) coincide when E is of finite dimension. It is evident that the following implications for convergence rates (c.r.) hold:

quadratic c.r. => superlinear c.r.
=> weakly superlinear c.r. => linear c.r.

The last implication follows from the fact that a weakly convergent sequence is bounded in the norm. In general, the reverse implications are false.

## 5.2 LINEAR AND QUADRATIC CONVERGENCE

We prove a sufficient condition for quadratic convergence.

THEOREM 5.3: Let F be Fréchet differentiable with Lipschitz continuous derivative in a neighborhood of some $\hat{u} \in H$ with $\nabla F(\hat{u}) = \Theta$ and let $\nabla^2 F(\hat{u})$ exist with continuous inverse. We consider the iteration

$$u_{i+1} = u_i - B_i^{-1} \nabla F(u_i) ,$$

where $\{B_i\}_{\mathbb{N}} \subset L(H)$, $B_i$ invertible, and

$$\|B_i - \nabla^2 F(\hat{u})\| \leq \nu \|u_i - \hat{u}\| \tag{5.5}$$

for all $i \in \mathbb{N}$.

If $u_0$ is in a neighborhood of $\hat{u}$, sufficiently small, then $\{u_i\}_{\mathbb{N}}$ converges to $\hat{u}$ and the rate of convergence is quadratic.

Proof: Let $\kappa$ be the Lipschitz constant of $\nabla F(u)$, and let $u_i$ lie in the neighborhood such that the Lipschitz continuity holds. Then

$$||B_i\ (u_{i+1} - \hat{u})|| = ||(B_i - \nabla^2 F(\hat{u}))(u_i - \hat{u})$$

$$- (\nabla F(u_i) - \nabla F(\hat{u}) - \nabla^2 F(\hat{u})(u_i - \hat{u}))|| \leq (\nu + \kappa)\ ||u_i - \hat{u}||^2 .$$

Since $\nabla^2 F(\hat{u})^{-1}$ is continuous there is some $\alpha > 0$ such that $||\nabla^2 F(\hat{u})v|| \geq \alpha\ ||v||$ for all $v \in H$,

$$||u_{i+1} - \hat{u}||\ (\alpha - \nu\ ||u_i - \hat{u}||\ )$$

$$\leq ||\nabla^2 F(\hat{u})(u_{i+1} - \hat{u})|| - ||(B_i - \nabla^2 F(\hat{u}))(u_{i+1} - \hat{u})||$$

$$\leq (\nu + \kappa)\ ||u_i - \hat{u}||^2$$

and assuming $||u_i - \hat{u}|| < \alpha\ \nu^{-1}$

$$||u_{i+1} - \hat{u}|| \leq \frac{(\nu + \kappa)\ ||u_i - \hat{u}||^2}{\alpha - \nu\ ||u_i - \hat{u}||} . \tag{5.6}$$

If $||u_i - \hat{u}|| \leq \varepsilon$ such that $\varepsilon \leq \frac{\alpha}{\nu + \kappa}$ and the Lipschitz condition holds, then from (5.6)

$$||u_{i+1} - \hat{u}|| \leq \frac{\nu + \kappa}{\alpha}\ ||u_i - \hat{u}||^2 \leq \varepsilon ,$$

and the same conditions hold for $||u_{i+1} - \hat{u}||$. Hence, if $||u_0 - \hat{u}|| \leq \varepsilon$, $\varepsilon$ determined as above, then the rate of convergence is quadratic.

THEOREM 5.4: Let F be Fréchet differentiable with Lipschitz continuous derivative in a neighborhood of some $\hat{u} \in H$ with $\nabla F(\hat{u}) = \Theta$ and let $\nabla^2 F(u)$ have continuous inverse. Let $\{B_i\}_{\mathbb{N}} \in H$ be a sequence of invertible operators and $\{u_i\}_{\mathbb{N}} \in H$ given by

$$u_{i+1} = u_i - B_i^{-1}\ \nabla F(u_i) .$$

Suppose that for some $\beta, \delta \in \mathbb{R}$ and all $i \in \mathbb{N}$

$$\|B_{i+1} - \nabla^2 F(\hat{u})\| \leq (1 + \beta\lambda_i) \|B_i - \nabla^2 F(\hat{u})\|$$

$$+ \delta (1 + \|B_i\|) \lambda_i , \tag{5.7}$$

$$\lambda_i = \max \{ \|u_{i+1} - \hat{u}\| , \|u_i - \hat{u}\| \} \quad \text{for all } i \in \mathbb{N} .$$

If $\|u_0 - \hat{u}\|$ and $\|B_0 - \nabla^2 F(\hat{u})\|$ are sufficiently small, $\{u_i\}_{\mathbb{N}}$ converges to $\hat{u}$ at a linear rate with $\nu < 1$ in (5.1).

For the proof we estimate the norms from both sides of the following identity

$$\nabla^2 F(\hat{u})(u_{i+1} - \hat{u}) + (B_i - \nabla^2 F(\hat{u}))(u_{i+1} - \hat{u})$$
$$= (B_i - \nabla^2 F(\hat{u}))(u_i - \hat{u}) - \nabla F(u_i) + \nabla F(\hat{u}) + \nabla^2 F(\hat{u})(u_i - \hat{u})$$

yielding for $\|u_i - \hat{u}\|$ small and some $\alpha > 0$

$$(\alpha - \|B_i - \nabla^2 F(\hat{u})\|) \|u_{i+1} - \hat{u}\|$$
$$\leq ( \kappa \|u_i - \hat{u}\| + \|B_i - \nabla^2 F(\hat{u})\| ) \|u_i - \hat{u}\| .$$

Since the latter inequality is similar to that in (9.16) and the proof of Theorem 9.9 is almost identical with that above, we refer the reader to Section 9.4.

## 5.3 CHARACTERIZATION OF SUPERLINEAR CONVERGENCE

First, we prove a short lemma in order to clarify the proofs.

LEMMA 5.5: Let F be Fréchet differentiable in a neighborhood of some $\hat{u} \in H$ and let $\nabla^2 F(\hat{u})$ exist with $\nabla^2 F(\hat{u})^{-1} \subset L(H)$, $\{u_i\}_{\mathbb{N}} \subset H$. Then

$$\lim_{i\to\infty} \frac{\|u_{i+1} - \hat{u}\|}{\|u_i - \hat{u}\|} = 0 \iff \lim_{i\to\infty} \frac{\|\nabla F(u_{i+1}) - \nabla F(\hat{u})\|}{\|u_{i+1} - u_i\|} = 0 . \tag{5.8}$$

Proof: Let $\{u_i\}_{\mathbb{N}}$ converge to $\hat{u}$ at a superlinear rate. Then $\{ \|u_{i+1} - u_i\| \, \|u_i - \hat{u}\|^{-1} \}_{\mathbb{N}}$ is bounded, because

$$\lim_{i\to\infty} \left| \frac{\|u_{i+1} - u_i\|}{\|u_i - \hat{u}\|} - 1 \right| \leq \lim_{i\to\infty} \frac{\|u_{i+1} - \hat{u}\|}{\|u_i - \hat{u}\|} = 0. \tag{5.9}$$

Hence for some $\alpha > 0$

$$\frac{\|\nabla F(u_{i+1}) - \nabla F(\hat{u})\|}{\|u_{i+1} - u_i\|} \leq \alpha \frac{\|u_{i+1} - \hat{u}\|}{\|u_i - \hat{u}\|} \frac{\|u_i - \hat{u}\|}{\|u_{i+1} - u_i\|},$$

which converges to zero because of (5.9) and the superlinear convergence.

If the statement on the right side of (5.8) holds, then by the continuity of $\nabla^2 F(\hat{u})^{-1}$ there is a $\beta > 0$ such that

$$0 = \lim_{i\to\infty} \frac{\|\nabla F(u_{i+1}) - \nabla F(\hat{u})\|}{\|u_{i+1} - u_i\|} \geq \beta \lim_{i\to\infty} \frac{\|u_{i+1} - \hat{u}\|}{\|u_{i+1} - u_i\|}$$

$$\geq \beta \lim_{i\to\infty} \left( 1 + \frac{\|u_i - \hat{u}\|}{\|u_{i+1} - \hat{u}\|} \right)^{-1},$$

yielding superlinear convergence.

We characterize superlinear convergence for a class of algorithms.

<u>THEOREM 5.6:</u> Let F be differentiable as in Lemma 5.5 at some point $\hat{u} \in H$, $\{B_i\}_{\mathbb{N}} \subset L(H)$, $\{u_i\}_{\mathbb{N}} \subset H$, $B_i$ invertible and

$$u_{i+1} = u_i - B_i^{-1} \nabla F(u_i) .$$

Let $\{u_i\}_{\mathbb{N}}$ converge to $\hat{u}$.

Then $\{u_i\}_{\mathbb{N}}$ converges at a superlinear rate to $\hat{u}$ and $\nabla F(\hat{u}) = \Theta$ if and only if

$$\lim_{i\to\infty} \frac{\|(B_i - \nabla^2 F(\hat{u}))(u_{i+1} - u_i)\|}{\|u_{i+1} - u_i\|} = 0 . \tag{5.10}$$

<u>Proof:</u> The following identity holds

$$(B_i - \nabla^2 F(\hat{u}))(u_{i+1} - u_i)$$

$$= \nabla F(u_{i+1}) - \nabla F(u_i) - \nabla^2 F(\hat{u})(u_{i+1} - u_i) - \nabla F(u_{i+1}). \quad (5.11)$$

By the smoothness assumptions on F and the convergence of $\{u_i\}_{\mathbb{N}}$ to $\hat{u}$, (5.10) is equivalent to

$$\lim_{i \to \infty} \frac{\| \nabla F(u_{i+1}) \|}{\| u_{i+1} - u_i \|} = 0 . \quad (5.12)$$

If (5.12) holds, then

$$\Theta = \lim_{i \to \infty} \nabla F(u_i) = \nabla F(\hat{u}) ,$$

hence (5.12) is equivalent to the right hand side of (5.8). Lemma 5.5 yields the statement to be proven.

Since in the infinite dimensional case (5.10) is hard to check for variable metric updates we shall characterize weak superlinear convergence.

<u>THEOREM 5.7:</u> Let F be twice continuously Fréchet differentiable in a neighborhood of $\hat{u} \in H$, $\nabla^2 F(\hat{u})^{-1} \in L(H)$, $\{u_i\}_{\mathbb{N}} \subset H$, $\{B_i\}_{\mathbb{N}} \subset L(H)$, $B_i$ invertible,

$$u_{i+1} = u_i - B_i^{-1} \nabla F(u_i) ,$$

and let $\{u_i\}_{\mathbb{N}}$ be linearly convergent to $\hat{u}$ at a rate $\nu < 1$. Then $\{u_i\}_{\mathbb{N}}$ converges at a weak superlinear rate to $\hat{u}$ and $\nabla F(\hat{u}) = \Theta$ if and only if for each $\ell \in H$

$$\lim_{i \to \infty} \frac{\langle \ell, (B_i - \nabla^2 F(\hat{u}))(u_{i+1} - u_i) \rangle}{\| u_{i+1} - u_i \|} = 0 . \quad (5.13)$$

*Proof:* Equation (5.11) and the convergence of $\{u_i\}_{\mathbb{N}}$ to $\hat{u}$ yield that (5.13) is equivalent to

$$\lim_{i \to \infty} \frac{\langle \ell, \nabla F(u_{i+1}) \rangle}{\| u_{i+1} - u_i \|} = 0 \qquad \text{for each } \ell \in H . \quad (5.14)$$

Since (5.14) implies $<\ell, \nabla F(\hat{u})> = 0$ for all $\ell \in H$, we obtain $\nabla F(\hat{u}) = 0$ from (5.14), which is equivalent to

$$\lim_{i \to \infty} \frac{|<\ell, \nabla F(u_{i+1}) - \nabla F(\hat{u})>|}{\|u_{i+1} - u_i\|} = 0 . \tag{5.15}$$

From the Fréchet differentiability we obtain

$$|<\ell, \nabla F(u_{i+1}) - \nabla F(\hat{u})>|$$

$$\geq |<\ell, \nabla^2 F(\hat{u})(u_{i+1} - \hat{u})>| - \|u_{i+1} - \hat{u}\| \varepsilon_i \tag{5.16}$$

where $\varepsilon_i \to 0$ for $i \to \infty$ .

Linear convergence of $\{u_i\}$ gives

$$\|u_{i+1} - u_i\| \leq (1 + \nu) \|u_i - \hat{u}\| . \tag{5.17}$$

(5.16) yields with (5.15) and (5.17)

$$\lim_{i \to \infty} \frac{|<\nabla^2 F(\hat{u}) \ell, u_{i+1} - \hat{u}>|}{\|u_i - \hat{u}\|} \leq \lim_{i \to \infty} \varepsilon_i \frac{\| u_{i+1} - \hat{u} \|}{u_i - \hat{u}} = 0.$$

This implies using the surjectivity of $\nabla^2 F(\hat{u})$ the weak convergence of $(u_{i+1} - \hat{u}) \|u_i - \hat{u}\|^{-1}$ to zero.

If on the other hand (5.4) holds, then

$$\lim_{i \to \infty} \frac{|<\ell, \nabla^2 F(\hat{u})(u_{i+1} - \hat{u})>|}{\|u_i - \hat{u}\|} = 0 \qquad \text{for each } \ell \in H .$$

Fréchet differentiability and boundedness of $(u_{i+1} - \hat{u}) \|u_i - \hat{u}\|^{-1}$ yield

$$0 = \lim_{i \to \infty} \left( \frac{|<\ell, \nabla F(u_{i+1}) - \nabla F(\hat{u})>|}{\|u_i - \hat{u}\|} + \varepsilon_i \frac{\|u_{i+1} - \hat{u}\|}{\|u_i - \hat{u}\|} \right)$$

$$= \lim_{i \to \infty} \frac{|<\ell, \nabla F(u_{i+1}) - \nabla F(\hat{u})>|}{\|u_{i+1} - u_i\|} \frac{\|u_{i+1} - u_i\|}{\|u_i - \hat{u}\|} . \tag{5.18}$$

Linear convergence implies

$$\frac{\|u_{i+1} - u_i\|}{\|u_i - \hat{u}\|} \geq 1 - \frac{\|u_{i+1} - \hat{u}\|}{\|u_i - \hat{u}\|} \geq 1 - \nu > 0 .$$

Hence (5.18) implies (5.15) which is equivalent to (5.4).

<u>REMARK:</u> If there exists some $\ell \in H$, $\varepsilon > 0$ such that

$$\frac{|\langle \ell, u_i - \hat{u}\rangle|}{\|u_i - \hat{u}\|} \geq \varepsilon \qquad \text{for all } i \in \mathbb{N} ,$$

then, under the assumptions of Theorem 5.7, superlinear convergence can be obtained.

For a short proof of this remark, note that (5.15) and (5.16) imply that

$$0 = \lim_{i \to \infty} \frac{\langle \ell, \nabla^2 F(\hat{u})(u_{i+1} - \hat{u})\rangle - \|u_{i+1} - \hat{u}\| \ \varepsilon_i}{\|u_{i+1} - \hat{u}\| + \|u_i - \hat{u}\|}$$

$$= \lim_{i \to \infty} \left( \frac{\langle \ell, \nabla^2 F(\hat{u})(u_{i+1} - \hat{u})\rangle}{\|u_{i+1} - \hat{u}\|} - \varepsilon_i \right) \left(1 + \frac{\|u_i - \hat{u}\|}{\|u_{i+1} - \hat{u}\|}\right)^{-1} .$$

Since the first term in the product does not tend to zero we obtain from the second term the superlinear convergence.
We have not made use of the linear convergence rate. Only the convergence of $\{u_i\}_{\mathbb{N}}$ to $\hat{u}$ was required.

## 5.4 WEAK SUPERLINEAR CONVERGENCE

In this section we prove the weak superlinear convergence of a class of updates. Let us consider updates of the form (3.22) with $\rho = 1$

$$\bar{B} = B + \langle b,p\rangle^{-1} \ (\,]y - Bp,b[\, + \,]b,y - Bp[\,)$$
$$+ \langle b,p\rangle^{-2} \ \langle p,y - Bp\rangle \ ]b,b[ \ . \qquad (5.19)$$

Let $\nabla^2 F(\hat{u})$ be strictly positive definite. Then multiplying $\bar{B}$ from left and right with

$$G = (\nabla^2 F(\hat{u}))^{-1/2} ,$$

(5.19) yields with $D = \nabla^2 F(\hat{u})$

$$G\bar{B}G - I = A^* \, (GBG - I) \, A$$

$$+ \langle b,p\rangle^{-1} \, ( \, ]G(y - Dp), Gb[ \; + \; ]Gb, y - Dp[ \; GA), \tag{5.21}$$

$$A = I - \langle b,p\rangle^{-1} \, ]G^{-1}p, Gb[ \; .$$

We assume that $b \in H$ is such that for some $\alpha_1 > 0$

$$\|Gb - G^{-1}p\| \le \alpha_1 \lambda \, \|G^{-1}p\| , \tag{5.22}$$

$$\lambda = \max \, ( \|\bar{u} - \hat{u}\| , \; \|u - \hat{u}\| ) \; .$$

<u>THEOREM 5.8:</u> Let F be twice Fréchet differentiable with Lipschitz continuous first derivative in a neighborhood of the optimal point $\hat{u} \in H$ and $\nabla^2 F(\hat{u})^{-1} \in L(H)$, $\nabla^2 F(\hat{u})$ self-adjoint strictly positive definite. Let $\{B_i\}_{\mathbb{N}}$ be a sequence of operators updated by (5.19) and let (5.22) hold. If $\|u_0 - \hat{u}\|$ and $\|B_0 - \nabla^2 F(\hat{u})\|$ are small enough, then $u_i$ converges to $\hat{u}$ at a weakly superlinear rate.

The proof follows the arguments of Lemma 4.1, 4.2, 5.1, 5.2 in [3]. The main difference is that we cannot use the Frobenius-norm $\|A\|_F = \operatorname{tr} A^T A$ of a matrix, but we use instead $\|A^* x\|$ for fixed $x \in H$.

Thus we obtain similar estimates as in [3]. Since that proof is rather technical and lengthy we omit the steps here, with the result,

$$\|(G\bar{B}G - I)^* \, x\| \le ((1 - \alpha\phi(x)^2)^{1/2} + \frac{5}{2} \, (1 - \beta)^{-1} \, \alpha_1 \lambda)$$

$$\cdot \; \|(GBG - I)^* \, x\| + \gamma \, \|x\| \; \|G\| \; \|y - Dp\| \; \|G^{-1}p\|^{-1} , \tag{5.23}$$

where $\alpha \in [\frac{3}{8} , 1]$, $\gamma > 0$ and

$$\phi(x) = \frac{\langle G^{-1}p, \, (GBG - I)^* x\rangle}{\|G^{-1}p\| \; \|(GBG - I)^* x\|} \; .$$

Taking the supremum over $\|x\|$ less or equal 1 yields

$$\|G\bar{B}G - I\| \le (1 + \beta_1\lambda) \; \|G\bar{B}G - I\| + \beta_2 \; \|y - Dp\| \quad (5.24)$$

for some $\beta_1, \beta_2 > 0$. Since for some $\beta_3$

$$\|y - \nabla^2 F(\hat{u})p\| \le \lambda \; \|p\| \le \max\,(\,\|\bar{u} - \hat{u}\|\,,\,\|u - \hat{u}\|\,)\,\|p\|$$

and G is bounded and strictly positive definite, (5.24) implies

$$\|\bar{B} - \nabla^2 F(\hat{u})\| \le (1 + \beta_1\beta_3\lambda) \; \|B - \nabla^2 F(\hat{u})\| + \beta_2\beta_3\lambda \,.$$

Therefore we can apply Theorem 5.4 which yields linear convergence.

For the proof of weak superlinear convergence we shall exploit the better estimate (5.23). Since

$$(1 - \alpha\phi(x)^2)^{1/2} \le 1 - \frac{\alpha}{2}\,\phi(x)^2$$

(5.23) yields

$$\frac{\alpha}{2}\,\phi(x)^2 \; \|(GBG - I)^* x\| \le \; \|(GBG - I)^* x\|$$

$$- \; \|(G\bar{B}G - I)^* x\| + \beta_4\lambda \,,$$

for some $\lambda > 0$. Writing this inequality with indices and summing both sides yields

$$\infty > \sum_{i=1}^{\infty} \phi_i(x)^2 \; \|(GB_iG - I)^* x\|$$

$$= \sum_{i=1}^{\infty} \frac{\langle G^{-1}p_i \,,\, (GB_iG - I)^* x\rangle^2}{\|G^{-1}p_i\|^2 \; \|(GB_iG - I)^* x\|} \,.$$

The numbers $\|(GB_iG - I)^* x\|$ are bounded from above by the linear convergence proof. Hence

$$\lim_{i\to\infty} \langle G^{-1}p_i \,,\, (GB_iG - I)^* x\rangle \, \|G^{-1}p_i\|^{-1} = 0$$

and

$$\lim_{i\to\infty} \frac{\langle x,\, (B_i - \nabla^2 F(\hat{u}))p_i\rangle}{\|p_i\|} = 0$$

for each $x \in H$, implying weak superlinear convergence by Theorem 5.7.

For the DFP update, $b = y$, condition (5.22) is satisfied because

$$\|Gy - G^{-1}p\| \leq \|G\| \ \|y - \nabla^2 F(\hat{u})p\| \leq \alpha_1 \lambda \ \|p\|$$

for some $\alpha_1 > 0$.

## 5.5 COMMENTS

Section 5.1: These are the usual definitions of Q-convergence with the possible exception of weak superlinear convergence (5.4).

Section 5.2: These theorems are standard statements for linear and quadratic convergence in $\mathbb{R}^n$.

Section 5.3: The characterization of superlinear convergence can be carried over to the Hilbert space case from $\mathbb{R}^n$ [1]. An analogous result holds for weak superlinear convergence assuming linear convergence.

Section 5.4: A class of updates and their superlinear convergence are treated in [1] and [2].

Several other results are found in [3 - 10], including conditions under which the step length can be set equal to unity and still obtain local convergence.

# 6 Conjugate gradient methods

## 6.1 MOTIVATION

Let $v_i$ be a descent direction at $u_i$, i.e. satisfying

$$<\nabla F(u_i), v_i> < 0 .$$

If an exact step length is determined, then

$$<\nabla F(u_i + \alpha_i v_i), v_i> = <\nabla F(u_{i+1}), v_i> = 0 . \qquad (6.1)$$

Choosing a descent direction at the next iteration point $u_{i+1}$, the negative gradient $-\nabla F(u_{i+1})$ can be corrected by $v_i$ without changing the descent property. (6.1) yields

$$<\nabla F(u_{i+1}), -\nabla F(u_{i+1}) + \beta_i v_i> = - \|\nabla F(u_{i+1})\|^2 < 0$$

for any $\beta_i > 0$. Denoting

$$v_{i+1} = -\nabla F(u_{i+1}) + \beta_i v_i , \qquad (6.2)$$

we have a sequence of descent vectors which are nonorthogonal to each other in the Euclidean inner product

$$<v_{i+1}, v_i> = -<\nabla F(u_{i+1}), v_i> + \beta_i \|v_i\|^2 = \beta_i \|v_i\|^2 \neq 0$$

in contrast to the steepest descent method for which

$$<\nabla F(u_{i+1}), \nabla F(u_i)> = 0 .$$

However, the parameter $\beta_i$ can be selected. It is well known that in the n-dimensional case for a quadratic objective

$$F(u) = \frac{1}{2} <u, Qu> + <b, u>$$

a "conjugacy" requirement of the descent directions

$$<v_{i+1}, Qv_i> = 0 \qquad (6.3)$$

and exact step length determination lead to termination of the algorithm within n steps. Therefore, we try to select $\beta_i$ such that (6.3) holds. In the quadratic case

$$\nabla F(u) = Qu$$

implies

$$0 = \langle v_{i+1}, Qv_i \rangle = -\langle Qu_{i+1}, Qv_i \rangle + \beta_i \langle v_i, Qv_i \rangle$$

and

$$\beta_i = \frac{\langle Qu_{i+1}, Qv_i \rangle}{\langle v_i, Qv_i \rangle} . \tag{6.4}$$

This formula can be rewritten using $u_{i+1} = u_i + \alpha_i v_i$

$$\beta_i = \frac{\langle Qu_{i+1}, Qu_{i+1} - Qu_i \rangle}{\langle v_i, Qu_{i+1} - Qu_i \rangle} \tag{6.5}$$

and with (6.1)

$$\beta_i = \frac{\|Qu_{i+1}\|^2 - \langle Qu_{i+1}, Qu_i \rangle}{- \langle v_i, Qu_i \rangle} . \tag{6.6}$$

We also have with (6.2), (6.3) and (6.1)

$$\langle Qu_{i+1}, Qu_i \rangle = - \langle Qu_{i+1}, v_i \rangle + \beta_{i-1} \langle Qu_{i+1}, v_{i-1} \rangle$$

$$= \beta_{i-1} \langle Qu_{i+1}, v_{i-1} \rangle = \beta_{i-1} \langle Qu_i, v_{i-1} \rangle - \beta_{i-1} \alpha_i \langle Qv_i, v_{i-1} \rangle = 0.$$

Hence (6.6) is identical with

$$\beta_i = \frac{\|Qu_{i+1}\|^2}{- \langle v_i, Qu_i \rangle} = \frac{\|Qu_{i+1}\|^2}{\|Qu_i\|^2 - \beta_i \langle v_{i-1}, Qu_i \rangle}$$

and

$$\beta_i = \frac{\|Qu_{i+1}\|^2}{\|Qu_i\|^2} . \tag{6.7}$$

## 6.2 SELECTION OF DESCENT DIRECTIONS

From the considerations for the quadratic case, we can deduce various descent directions by remembering that $Qu_{i+1} = \nabla F(u_{i+1})$. The most compact form (6.7) yields the formula of Fletcher and Reeves,

$$\beta_i = \frac{\|\nabla F(u_{i+1})\|^2}{\|\nabla F(u_i)\|^2} \quad .$$

From all the other possibilities let us mention the parameter $\beta_i$ proposed by Wolfe which can be deduced from (6.5)

$$\beta_i = \frac{\langle \nabla F(u_{i+1}),\ \nabla F(u_{i+1}) - \nabla F(u_i)\rangle}{\langle v_i,\ \nabla F(u_{i+1}) - \nabla F(u_i)\rangle} \quad . \tag{6.8}$$

Hence the conjugate direction algorithm is of the following form:

Step 0. Select $u_0 \in H,\ \ v_0 = -\nabla F(u_0),\ \ i = 0$ .

Step 1. Determine $\alpha_i$ by some step length rule.

Step 2. Set

$$u_{i+1} = u_i + \alpha_i v_i \ .$$

Step 3. Set

$$v_{i+1} = -\nabla F(u_{i+1}) + \beta_i v_i$$

for some correction formula for $\beta_i$.

Step 4. Increase i by 1 and go to Step 1.

It has been very important for the motivation via conjugacy to use the exact step length rule. However, this rule is quite inconvenient for implementation of an algorithm.

In Chapter 4 we have developed an approach based on inexact step length rules when the descent is given by the image of the negative gradient under a linear map. With the definitions,

$$p_i = u_{i+1} - u_i, \tag{6.9}$$

$$y_i = \nabla F(u_{i+1}) - \nabla F(u_i), \tag{6.10}$$

Wolfe's update (6.8) is given by

$$v_{i+1} = -\nabla F(u_{i+1}) + \frac{\langle \nabla F(u_{i+1}), y_i \rangle}{\langle v_i, y_i \rangle} v_i$$

$$= -\nabla F(u_{i+1}) + \frac{\langle \nabla F(u_{i+1}), y_i \rangle}{\langle p_i, y_i \rangle} p_i$$

$$= (I - \frac{]p_i, y_i[}{\langle p_i, y_i \rangle})(-\nabla F(u_{i+1})) . \qquad (6.11)$$

This form is reminiscent of the variable metric update

$$v_{i+1} = -B_{i+1}^{-1} \nabla F(u_{i+1}) .$$

However, the update formula for $B_{i+1}$ has a term $B_i$ which is always set to the identity,

$$B_i = I \qquad (6.12)$$

in (6.11). Hence, we investigate some variable metric updates using the condition (6.12). The most general form of self-adjoint rank-two updates satisfying an equation is given by (3.17). Hence we require that B satisfies the quasi-Newton equation (3.15)

$$B_{i+1} p_i = y_i ,$$

however, with $B_i = I$.

Equation (3.22) for example then yields

$$\bar{B} = \rho(I + \frac{\langle p,p \rangle}{\langle b,p \rangle^2} ]b,b[ - \frac{1}{\langle b,p \rangle} (]p,b[ + ]b,p[))$$

$$- \frac{\langle p,y \rangle}{\langle b,p \rangle^2} ]b,b[ + \frac{1}{\langle b,p \rangle} (]y,b[ + ]b,y[) , \qquad (6.13)$$

where $b \in H$ and $\rho > 0$ are parameters.

## 6.3 GLOBAL CONVERGENCE

In order to prove global convergence for algorithms with an inexact step length rule let us apply Theorem 4.2. There are two requirements which have to be met, the positive definiteness of $\bar{B}$ and the uniform boundedness of the condition numbers.

Formula (6.13) suggests two main choices for b

$$b = p \quad \text{or} \quad b = y .$$

The first theorem deals with the case b = y.

Assume that Assumption $A_2$ of Section 2.1 holds.

<u>THEOREM 6.1:</u> If $\{u_i\}_{\mathbb{N}}$ is a sequence constructed by a conjugate gradient method using a descent determination by

$$v = - \bar{B}^{-1} \nabla F(u) ,$$

$$\bar{B} = \rho (I + \frac{<p,p>}{<p,y>^2} ]y,y[ - \frac{1}{<p,y>} (]p,y[ + ]y,p[))$$

$$+ \frac{1}{<p,y>} ]y,y[ , \qquad \varepsilon_2 \geq \rho \geq \varepsilon_1 > 0 ,$$

and a step length rule which yields the Zoutendijk condition (2.2), then $\{u_i\}$ converges to the optimal point $\hat{u}$.

<u>Proof:</u> First, we show the positive definiteness of $\bar{B}$. Since

$$\rho_1 = <p,y>^{-2} > 0 ,$$

$\bar{B}$ is positive definite if (3.39) holds, i.e.

$$<y,y>^2 - <y,y> (<y,y> - \rho <p,y>)$$

$$= \rho <y,y> \ <p,y> \ > \ 0 .$$

The boundedness of the condition numbers is proved by Lemma 4.3. $\Lambda$ and $\Delta$ in (3.37) and (3.38) are given by

$$\Delta = \rho^2 <p,y>^{-2} <p,p> <y,y> - \rho^2 ,$$

$$\Lambda = \frac{1}{2} <p,y>^{-2} (<p,y> <y,y> - 2\rho <p,y>^2 + \rho <p,p> <y,y>).$$

If $<p,y> > \rho <p,p>$, then (4.7) holds and

$$\text{cond } \bar{B} \leq (\rho + \Lambda + (\Delta + \Lambda^2)^{1/2})^2 (\rho^2 + 2\rho\Lambda - \Delta)^{-1} .$$

A short calculation shows

$$\rho^2 + 2\rho\Lambda - \Delta = \rho \ \frac{\langle y,y\rangle}{\langle p,y\rangle} ,$$

which is by the assumption on $\rho$ and by Chapter 1.4 uniformly bounded from zero. On the other hand, the terms in $\Delta$ and $\Lambda$ are bounded from above. Thus the condition numbers do not tend to infinity for the case in which 4.9 applies.

In particular, we have shown the convergence for the conjugate gradient method which is deduced from the scaled DFP update:

$$\overline{B} = \frac{\langle p,y\rangle}{\langle p,p\rangle} \ I + \frac{2}{\langle p,y\rangle} \ ]y,y[ \ - \frac{1}{\langle p,y\rangle} \ (]p,y[ \ + \ ]y,p[) \ .$$

Let us consider now $b = p$ in (6.13), which is based on the PSB update.

<u>THEOREM 6.2:</u> Let the assumptions of Theorem 6.1 hold with

$$\overline{B} = \rho(I - \frac{1}{\langle p,p\rangle} \ ]p,p[) - \frac{\langle p,y\rangle}{\langle p,p\rangle^2} \ ]p,p[ \ + \frac{1}{\langle p,p\rangle} \ (]y,p[ \ + \ ]p,y[)$$

Let $\rho$ satisfy for some $\varepsilon_1 > 0$

$$\varepsilon_2 \geq \rho \geq \langle p,y\rangle^{-1} \ \langle p,p\rangle^{-1} \ (\langle p,y\rangle^2 - \langle p,p\rangle \ \langle y,y\rangle)$$
$$+ \ \varepsilon_1 \ \langle p,p\rangle^2 \ . \qquad (6.14)$$

If the step size rule yields Zoutendijk's condition, then global convergence holds.

<u>Proof:</u> Concerning positive definiteness we apply Corollary 3.11, i.e. (3.39). The number

$$\langle p,y\rangle^2 - \langle p,p\rangle \ (\langle y,y\rangle - \rho\langle p,y\rangle)$$

is nonnegative if (6.14) holds.

For $\Lambda$ and $\Delta$ in (3.37) and (3.38) we calculate

$$\Lambda = \frac{1}{2} \ ( \ \frac{\langle y,y\rangle}{\langle p,p\rangle} - \rho) \ ,$$

$$\Delta = \frac{1}{\langle p,p\rangle^2} \ (\langle y,y\rangle \ \langle p,p\rangle - \langle p,y\rangle^2) \ .$$

With (6.14) we obtain

$$\rho^2 + 2\rho\Lambda - \Delta = \langle p,p\rangle^{-2} \;(\rho\langle p,y\rangle \langle p,p\rangle + \langle y,y\rangle \langle p,p\rangle - \langle p,y\rangle^2) \geq \varepsilon_1 .$$

As in the preceding proof, we obtain the boundedness of $\Lambda$ and $\Delta$ by Assumption A2 and Chapter 1.4. Hence the condition numbers are bounded.

As a final example, let us treat the BFGS version where

$$\rho = \frac{\langle p,y\rangle}{\langle p,p\rangle} , \quad \zeta = \frac{\langle p,p\rangle^2}{\langle p,y\rangle} , \quad b = p , \quad z = y - \rho p ,$$

$$\zeta_1 = \langle p,p\rangle^{-2}$$

<u>THEOREM 6.3:</u> Under the same assumption as in the preceding theorems with

$$\overline{B} = \frac{\langle p,y\rangle}{\langle p,p\rangle} \left(I - \frac{]p,p[}{\langle p,p\rangle}\right) + \frac{]y,y[}{\langle p,y\rangle}$$

we obtain global convergence.

<u>Proof:</u> The proof is the same as before with the results

$$\Lambda = \frac{1}{2} \left( \frac{\langle y,y\rangle}{\langle p,y\rangle} - \frac{\langle p,y\rangle}{\langle p,p\rangle} \right) ,$$

$$\Delta = \frac{\langle y,y\rangle}{\langle p,p\rangle} - \frac{\langle p,y\rangle^2}{\langle p,p\rangle^2} ,$$

$$\rho^2 + 2\rho\Lambda - \Delta = \frac{\langle p,y\rangle^2}{\langle p,p\rangle^2} .$$

For an implementation we need the inverse operators $B^{-1}$, but they are easy to compute with the results of Section 3.3.

## 6.4 COMMENTS

<u>Section 6.1:</u> The conjugate gradient method was suggested by Hestenes and Stiefel [1].

Section 6.2: The formula of Fletcher and Reeves is given in [2]. For Wolfe's update see [3], p. 318. So-called memoryless quasi-Newton methods are presented in [4] with several numerical tests and in [5] with a convergence proof for an inexact step size rule. In [6] we find further results on the relationship between variable metric and conjugate gradient methods.

Section 6.3: Global convergence has been proved in [5] for a particular step size rule. In this regard see also [3]. Convergence proofs for general classes of conjugate gradient methods are outlined in [7]. See [8] for various inexact step size rules and convergence results in connection with conjugate gradient methods. Daniel [9], [10] and Fortuna [11], [12] investigate convergence and convergence rate in Hilbert spaces.

# 7 Conditional gradient methods

## 7.1 MOTIVATION

We now consider constrained problems of the following type:

Let H be a Hilbert space and U a convex, closed, bounded subset of H, called the feasible set,

$$F : U \to \mathbb{R}$$

the objective functional, which is Fréchet differentiable on an open subset of H containing U.

The constrained optimization problem consists of finding an element $\hat{u} \in U$ such that

$$F(\hat{u}) \le F(u) \qquad \text{for all } u \in U . \tag{7.1}$$

The idea of the conditional gradient method is to expand F at an iteration point $u_i \in U$ in a Taylor series up to first order,

$$F_1(u) = F(u_i) + \langle \nabla F(u_i), u - u_i \rangle \quad . \tag{7.2}$$

A new descent direction is obtained by minimizing $F_1(u)$ on the feasible set U:

Find $\bar{u}_i \in U$ such that

$$F_1(\bar{u}_i) \le F_1(u) \qquad \text{for all } u \in U$$

or equivalently

$$\langle \nabla F(u_i), \bar{u}_i - u_i \rangle \le \langle \nabla F(u_i), u - u_i \rangle \quad \text{for all } u \in U. \tag{7.3}$$

If $u_i \in U$ is optimal, i.e. it satisfies (7.1), then

$$\langle \nabla F(u_i), u - u_i \rangle \ge 0 \qquad \text{for all } u \in U, \tag{7.4}$$

i.e. $\bar{u}_i = u_i$ solves (7.3). Assuming F to be convex, then (7.4) is also equivalent with the optimality of $u_i$. This implies that $\bar{u}_i = u_i$ is characteristic for optimality.

If

$$\langle \nabla F(u_i), \bar{u}_i - u_i \rangle < 0 ,$$

then

$$v_i = \bar{u}_i - u_i \tag{7.5}$$

is a descent direction and we determine $u_{i+1}$ with a line search, for example:

Find $\lambda_i \in [0,1]$ such that

$$F(u_i + \lambda_i(\bar{u}_i - u_i)) \le F(u_i + \lambda(\bar{u}_i - u_i))$$
$$\text{for all } \lambda \in [0,1] \ .$$

The restriction of the step length to the interval [0,1] yields the feasibility of

$$u_{i+1} = u_i + \lambda_i(\bar{u}_i - u_i) \in U \ . \tag{7.6}$$

Let us summarize the prototype of a conditional gradient method:

<u>ALGORITHM 7A:</u>

Step 0. Select $u_0 \in U$.

Step 1. Determine $\bar{u}_i \in U$ such that for all $u \in U$

$$<\nabla F(u_i), \bar{u}_i> \le <\nabla F(u_i), u> \ .$$

Step 2. Define

$$v_i = \bar{u}_i - u_i$$

and determine $\lambda_i$ by some step length procedure.

Step 3. Define

$$u_{i+1} = u_i + \lambda_i v_i \ ,$$

set $i = i + 1$ and go to step 1.

Step 1 is well defined because U is a weakly compact set and $<\nabla F(u_i), u>$ a linear continuous functional. If $v_i \neq \Theta$, then it is also possible to achieve descent, i.e.

$$F(u_{i+1}) < F(u_i) \ ,$$

when $\lambda_i$ is chosen small enough, because

$$<\nabla F(u_i), v_i> < 0 \ .$$

## 7.2 GLOBAL CONVERGENCE

Although we can modify many of the exact step size rules of Chapter 2 to our constrained problem, we have to use a different condition than the Zoutendijk condition in order to prove convergence.

<u>THEOREM 7.1:</u> Let $F$ be convex on $U \subset H$, $\{u_i\}_{\mathbb{N}}$, $\{u_i\}_{\mathbb{N}} \subset U$, such that for each $i \in \mathbb{N}$

$$\langle \nabla F(u_i), v_i \rangle \le \langle \nabla F(u_i), u - u_i \rangle \quad \text{for all } u \in U . \tag{7.7}$$

If

$$\lim_{i \to \infty} \langle \nabla F(u_i), v_i \rangle = 0 \tag{7.8}$$

holds, then

$$\limsup_{i \to \infty} F(u_i) \le F(u) \quad \text{for all } u \in U . \tag{7.9}$$

<u>Proof:</u> Let $u \in U$ be an arbitrary, feasible element. Then by the convexity of $F$ and (7.7),

$$\langle \nabla F(u_i), v_i \rangle \le \langle \nabla F(u_i), u - u_i \rangle \le F(u) - F(u_i) .$$

Taking the limit for $i \to \infty$, we obtain (7.9).

Therefore, if $F$ is even weakly continuous, there exist weak accumulation points of $\{u_i\}_{\mathbb{N}}$ each of which is an optimal point by Theorem 7.1. Let us investigate several step length rules and prove condition (7.8).

<u>LEMMA 7.2:</u> Let $F$ have a Lipschitz continuous derivative on $U$ and let $\alpha_i$ in Algorithm 7A be determined by the <u>exact step length rule:</u>

Find $\alpha_i \in [0,1]$ such that

$$F(u_i + \alpha_i v_i) \le F(u_i + \alpha v_i) \quad \text{for all } \alpha \in [0,1]. \tag{7.10}$$

Then condition (7.8) holds.

<u>Proof:</u> Since $v_i$ are descent directions we have

$$F(u_{i+1}) < F(u_i) \quad \text{for all } i \in \mathbb{N} .$$

Since U is bounded the sequence of function values $\{F(u_i)\}_{\mathbb{N}}$ is convergent.

Using (7.10) with the Taylor expansion formula gives

$$F(u_{i+1}) - F(u_i) \le F(u_i + \alpha v_i) - F(u_i)$$

$$\le \alpha <\nabla F(u_i), v_i> + \alpha^2 \frac{k}{2} \|v_i\|^2 \qquad (7.11)$$

for all $\alpha \in [0,1]$, $i \in \mathbb{N}$. Since $\{F(u_i)\}_{\mathbb{N}}$ converges, (7.11) yields for any $\alpha \in [0,1]$

$$\nu = \sup_{u \in U} \|u\| ,$$

$$- \alpha^2 \frac{k}{2} \nu^2 \le \liminf_{i \to \infty} <\nabla F(u_i), v_i> \le 0 .$$

Because $\alpha > 0$ is arbitrary, this implies (7.8).

<u>LEMMA 7.3:</u> Let F be twice Fréchet differentiable and satisfy for some m, M > 0

$$m \|u\|^2 \le <u, \nabla^2 F(v)(u)> \le M \|u\|^2$$

for all $u \in H$, $v \in U$.

Consider a sequence $\{u_i\}_{\mathbb{N}} \subset U$ which has been constructed by Algorithm 7A with one of the following step length rules:

<u>1) Goldstein's step length rule</u>

Let $\gamma \in (0.5,1)$ be given. If

$$F(u_i + v_i) - F(u_i) < (1 - \gamma) <\nabla F(u_i), v_i> , \qquad (7.12)$$

then set $\alpha_i = 1$. Otherwise select $\alpha_i \in (0,1]$ such that

$$\alpha_i (1 - \gamma) <\nabla F(u_i), v_i> \le F(u_i + \alpha_i v_i) - F(u_i)$$

$$\le \alpha_i \gamma <\nabla F(u_i), v_i> \qquad (7.13)$$

holds.

<u>2) Powell's step length rule</u>

Let $\gamma, \mu \in \mathbb{R}$ with $0 < \gamma \le \mu < 1$ be given. If

$$<\nabla F(u_i + v_i), v_i> < \mu <\nabla F(u_i), v_i> , \qquad (7.14)$$

then set $\alpha_i = 1$. Otherwise choose $\alpha_i \in (0,1]$ such that

$$F(u_i + \alpha_i v_i) - F(u_i) \le \gamma\alpha_i \langle \nabla F(u_i), v_i \rangle \tag{7.15}$$

and

$$\mu \langle \nabla F(u_i), v_i \rangle \le \langle \nabla F(u_i + \alpha_i v_i), v_i \rangle \tag{7.16}$$

Then the condition (7.8) is true.

Proof: If $\alpha_i$ can be chosen from the interval (0,1], then we have by the proof of Theorem 2.3 and 2.4, that for some constant $\beta > 0$

$$F(u_{i+1}) - F(u_i) \le \beta \langle \nabla F(u_i), v_i \rangle < 0 . \tag{7.17}$$

If the cases (7.12) or (7.14) apply, then (7.17) also holds with a different constant $\beta$. However, these constants are independent of i and the convergence of $\{F(u_i)\}_{\mathbb{N}}$ gives with (7.17) the convergence of $\{\langle \nabla F(u_i), v_i \rangle\}_{\mathbb{N}}$ to zero.

The preceding has dealt with step length rules which require a finite iteration. If we have knowledge of the Lipschitz constant of the first derivative, then we can use a rule which does not require an iterative process.

LEMMA 7.4: Let F be Fréchet differentiable such that for some $k > 0$

$$\|\nabla F(u) - \nabla F(v)\| \le k \|u - v\| \qquad \text{for all } u, v \in U .$$

If $\{u_i\}$ is a sequence constructed by Algorithm 7A with the step length rule

$$\alpha_i = \min \{1, -\beta_i \langle \nabla F(u_i), v_i \rangle \|v_i\|^{-2} \} \tag{7.18}$$

such that for some $\varepsilon_1, \varepsilon_2 > 0$

$$\beta_i \in [\varepsilon_1, (2 - \varepsilon_2) k^{-1}] \tag{7.19}$$

for all $i \in \mathbb{N}$, then (7.8) is satisfied.

Proof: By assumption,

$$F(u_{i+1}) - F(u_i) \le \alpha_i \langle \nabla F(u_i) v_i \rangle + \alpha_i^2 \frac{k}{2} \|v_i\|^2 . \tag{7.20}$$

If $\alpha_i = -\beta_i <\nabla F(u_i), v_i> \|v_i\|^{-2} \geq -\varepsilon_1 <\nabla F(u_i), v_i> \|v_i\|^{-2}$,

then we obtain as in the proofs of Theorem 2.3 and 2.4

$$F(u_{i+1}) - F(u_i) \leq -\gamma <\nabla F(u_i), v_i>^2 \|v_i\|^{-2} \leq 0 \tag{7.21}$$

for some $\gamma > 0$.

If

$$\alpha_i = 1 < -\beta_i <\nabla F(u_i), v_i> \|v_i\|^{-2},$$

then (7.20) implies

$$F(u_{i+1}) - F(u_i) \leq <\nabla F(u_i), v_i> (1 - \beta_i \frac{k}{2})$$

$$\leq <\nabla F(u_i), v_i> \frac{\varepsilon_2}{2} \leq 0 . \tag{7.22}$$

Since $\{F(u_i)\}_{\mathbb{N}}$ is convergent, (7.21) and (7.22) imply (7.8). Suppose we do not know the Lipschitz constant explicitly. Then there is also a rule for selecting step sizes such that convergence is achieved.

LEMMA 7.5: Let F be convex and Fréchet differentiable with continuous Lipschitz derivative on U.

Let $\{u_i\}_{\mathbb{N}}$ be a sequence of iterates which are constructed by Algorithm 7A and with step sizes $\{\alpha_i\}_{\mathbb{N}} \subset (0,1]$ such that

$$\sum_{i=1}^{\infty} \alpha_i = \infty , \qquad \sum_{i=1}^{\infty} \alpha_i^2 < \infty \tag{7.23}$$

Then $\lim_{i \to \infty} F(u_i) = \min_{u \in U} F(u) = F(\hat{u})$.

Proof: The minimum of F on U is attained, since F is convex and U weakly compact.

As usual we exploit the inequality

$$F(u_{i+1}) - F(u_i) \leq \alpha_i <\nabla F(u_i), v_i> + \alpha_i^2 \frac{k}{2} \|v_i\|^2 .$$

Defining

$$\rho_i = F(u_i) - F(\hat{u}), \qquad \nu = \sup_{u \in U} \|u\|$$

$$\rho_{i+1} \leq \rho_i + \alpha_i <\nabla F(u_i), v_i> + \alpha_i^2 \frac{k}{2} \nu^2$$

$$\leq \rho_i + \alpha_i <\nabla F(u_i), \hat{u} - u_i> + \alpha_i^2 \frac{k}{2} \nu^2$$

$$\leq (1 - \alpha_i)\, \rho_i + \alpha_i^2 \frac{k}{2} \nu^2 \tag{7.24}$$

using step 1 in Alg. 7A and the convexity of F.

Rewriting (7.24) as

$$\rho_{i+1} - \rho_i \leq - \alpha_i \rho_i + \alpha_i^2 \frac{k}{2} \nu^2 \quad \text{for all } i \in \mathbb{N}, \tag{7.25}$$

we obtain by summing both sides and using (7.23) that

$$- \infty < \liminf_{i \to \infty} (\rho_{i+1} - \rho_i) - \frac{k}{2} \nu^2 \sum_{i=1}^{\infty} \alpha_i^2$$

$$\leq - \sum_{i=1}^{\infty} \alpha_i \rho_i . \tag{7.26}$$

Since the sum of the $\{\alpha_i\}_{\mathbb{N}}$ is infinite, (7.26) implies

$$\liminf_{i \to \infty} \rho_i = 0 . \tag{7.27}$$

Hence there exist a subsequence $\Lambda \subset \mathbb{N}$ such that $\{\rho_i\}_\Lambda$ converges to zero.

For each $\varepsilon > 0$, there exists $j_o \in \Lambda$ such that

$$\rho_{j_o} \leq \varepsilon \quad \text{and} \quad \sum_{i=j_o}^{\infty} \alpha_i^2 \leq \varepsilon .$$

(7.25) yields for each $n \in \mathbb{N}$

$$0 \leq \rho_{j_o+n} \leq \rho_{j_o} + \frac{k}{2} \nu^2 \sum_{i=j_o}^{\infty} \alpha_i^2 \leq \varepsilon \left(1 + \frac{k}{2} \nu^2\right) .$$

Therefore we have

$$\lim_{i \to \infty} \rho_i = 0 .$$

## 7.3 CONVERGENCE RATES

Since the conditional gradient method only uses the Taylor expansion of F in terms of first order, we can expect at best a linear convergence rate. We shall investigate the convergence rate of the functional values.

DEFINITION 7.6: A convex subset U of a Hilbert space H is called strongly convex if there exists $\gamma > 0$ such that

$$u, v \in U, \quad w \in H, \quad \|w\| \le \gamma \ \|u - v\|^2$$

implies

$$w + \frac{1}{2}(u + v) \in H .$$

<u>THEOREM 7.7:</u> Let F be convex and Fréchet differentiable with Lipschitz continuous derivative on U, U strongly convex.

If $\{u_i\}_{\mathbb{N}}$ is a sequence which is generated by Algorithm 7A with a step size rule by Goldstein, Powell or (7.18), such that $\|\nabla F(u_i)\| \ge \delta > 0$ for some $\delta > 0$ and all $i \in \mathbb{N}$ large enough, then there is a $\nu \in (0,1)$ such that

$$|F(u_{i+1}) - F(\hat{u})| \le \nu \ |F(u_i) - F(\hat{u})|$$

for all $i \in \mathbb{N}$ large enough.

<u>Proof:</u> Consider the identity

$$<\nabla F(u_i), v_i> = 2<\nabla F(u_i), \frac{1}{2}(u_i + \bar{u}_i)$$

$$- \gamma \ \nabla F(u_i) \ \|\nabla F(u_i)\|^{-1} \ \|u_i - \bar{u}_i\|^2 >$$

$$+ 2<\nabla F(u_i), \gamma \ \nabla F(u_i) \ \|\nabla F(u_i)\|^{-1} \ \|u_i - \bar{u}_i\|^2 - u_i> .$$

The long expression in the scalar product of the first term on the right hand side is an element of U, because U is strongly convex. Therefore, step 1 of Algorithm 7A yields

$$<\nabla F(u_i), v_i> \ge 2<\nabla F(u_i), v_i> + 2\gamma \ \|\nabla F(u_i)\| \ \|v_i\|^2$$

and for $i \ge i_0$, $i_0 \in \mathbb{N}$ fixed,

$$- <\nabla F(u_i), v_i> \ge 2\delta\gamma \ \|v_i\|^2 \ . \tag{7.28}$$

Reading the proofs of Lemmas 7.3 and 7.4 we see that for each $i \in \mathbb{N}$ there is a constant $\sigma > 0$ such that either

$$F(u_{i+1}) - F(u_i) \le \sigma <\nabla F(u_i), v_i> \tag{7.29}$$

or

$$F(u_{i+1}) - F(u_i) \le \sigma <\nabla F(u_i), v_i>^2 \ \|v_i\|^{-2} ,$$

which gives with (7.28)

$$F(u_{i+1}) - F(u_i) \le \frac{\sigma}{2\delta\gamma} \ <\nabla F(u_i), v_i> .$$

In any case (7.29) is true for all $i \geq i_o$ with an appropriate $\sigma > 0$. We suppose that $\sigma \in (0,1)$.

Hence by the convexity of F

$$F(u_{i+1}) - F(u_i) \leq \sigma \, (F(\hat{u}) - F(u_i))$$

and

$$F(u_{i+1}) - F(\hat{u}) \leq (1 - \sigma) \, (F(u_i) - F(\hat{u})), \quad i \geq i_o .$$

The condition

$$\|\nabla F(u_i)\| \geq \delta > 0 \qquad \text{for } i \geq i_o ,$$

is satisfied if $\{u_i\}_{\mathbb{N}}$ converges to the optimal point $\hat{u}$ and $\nabla F(u) \neq \Theta$. These are reasonable assumptions because we are interested in convergence rates if convergence itself is assured. The assumption that the gradient of the objective functional at the optimal point does not vanish indicates that this problem is unconstrained.

If we do not have the strong convexity property of the feasible set, then a slower convergence rate is obtained.

THEOREM 7.8: Let F be convex and Fréchet differentiable with Lipschitz continuous derivative on U. If $\{u_i\}_{\mathbb{N}}$ is a sequence which is generated by Algorithm 7A and with a step size rule by Goldstein, Powell or (7.18), then there is a constant $\zeta > 0$ and an index $i_o \in \mathbb{N}$ such that for $i \geq i_o$

$$|F(u_i) - \inf_{u \in U} F(u)| \leq \frac{\zeta}{i}$$

Proof: Similarly as in the proof of Theorem 7.7 we remark that for each $i \in \mathbb{N}$ there is a constant $\sigma > 0$ such that either

$$F(u_{i+1}) - F(u_i) \leq \sigma \langle \nabla F(u_i), v_i \rangle$$

or

$$F(u_{i+1}) - F(u_i) \leq \sigma \langle \nabla F(u_i), v_i \rangle^2 \, \|v_i\|^{-2} .$$

Hence

$$- \langle \nabla F(u_i), v_i \rangle$$

$$\leq \max \left(-\frac{1}{\sigma}(F(u_{i+1})-F(u_i)), \left(-\frac{\nu^2}{\sigma}(F(u_{i+1})-F(u_i))\right)^{1/2}\right), \qquad (7.30)$$

$$\nu = \sup_{u \in U} \|u\| .$$

Since $\{F(u_i)\}_{\mathbb{N}}$ decreases, $\{F(u_{i+1})-F(u_i)\}_{\mathbb{N}}$ converges to zero.

Let $u \in U$ be arbitrary and define

$$\rho_i = F(u_i) - \inf_{u \in U} F(u) .$$

Then (7.30) implies that for some constant $\sigma_1$ and $i_o \in \mathbb{N}$

$$- \langle \nabla F(u_i), v_i \rangle \leq \sigma_1 (\rho_i - \rho_{i+1})^{1/2} . \qquad (7.31)$$

Convexity of F and step 1 of Algorithm 7A yield

$$\langle \nabla F(u_i), v_i \rangle \leq \langle \nabla F(u_i), u - u_i \rangle \leq F(u) - F(u_i) , \qquad (7.32)$$

and (7.31) and (7.32) together give

$$\rho_i^2 \leq \sigma_1^2 (\rho_i - \rho_{i+1}) . \qquad (7.33)$$

Define for $i \geq i_o$

$$\tau_i = i\rho_i .$$

Dividing (7.33) by $\rho_i$ we obtain

$$\frac{\tau_i}{i} \leq \sigma_1^2 \left(1 - \frac{\tau_{i+1}}{\tau_i} \frac{i}{i+1}\right) , \qquad (7.34)$$

where the term in parentheses has to be positive because $\tau_i > 0$.

If $\sigma_1^2 < \tau_i$, then (7.34) implies

$$\frac{\tau_i}{i} \leq \tau_i \left(1 - \frac{\tau_{i+1}}{\tau_i} \frac{i}{i+1}\right)$$

and

$$\frac{\tau_{i+1}}{\tau_i} \leq \frac{i+1}{i} \frac{i-1}{i} < 1 ,$$

$$\tau_{i+1} < \tau_i .$$

Otherwise,

$$\tau_i \leq \sigma_1^2 \qquad (7.35)$$

holds. In either case $\tau_i$ is bounded by some $\zeta > 0$ and hence

$$\rho_i \leq \frac{\zeta}{i} \qquad \text{for all } i \geq i_o .$$

## 7.4 A METHOD BASED ON THE BANG-BANG PRINCIPLE

For certain optimal control problems it is possible to prove that the optimal control satisfies

$$|\hat{u}(t)| = 1 \qquad \text{for all } t \in [0,T] ,$$

except at finitely many points, where it switches from -1 to +1. For a discussion of this property see Section 11.1. This principle gives a motivation for the method discussed in this chapter.

The determination of $\bar{u}_i$ in step 1 of Algorithm 7A for control problems with

$$U = \{u \in L_\infty [0,T] : |u(t)| \leq 1 \quad \text{a.e. in } [0,T]\}$$

is rather simple because

$$\int_0^T g_i(t) \; (\bar{u}_i(t) - u(t))dt \leq 0 \qquad \text{for all } u \in U$$

with $g_i = \nabla F(u_i)$ holds if and only if

$$\bar{u}_i(t) = - \operatorname{sgn} g_i(t)$$

for almost all $t \in [0,T]$ such that $g_i(t) \neq 0$. If $g_i$ has only finitely many zeroes in $[0,T]$, then $\bar{u}_i$ is a bang-bang control. However in step 3 of the conditional gradient algorithm we determine a control $u_{i+1}$ which is a convex combination of two controls and does not yield in general a function which is of bang-bang type, cf. Fig. 7.1.

If we start with a bang-bang control $u_o$ and if $\bar{u}_o$ is also bang-bang, then we construct new controls as convex combinations of the switching points of $\bar{u}_o$ and $u_o$.

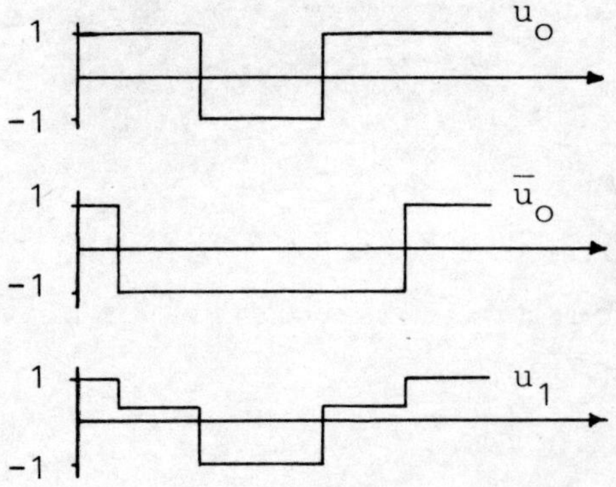


Figure 7.1: Conditional Gradient Method

Algorithm 7B

Step 0: Set $u_0 \equiv 1$, $n = 1$, $i = 0$.

Step 1: Let $u_i$ be a bang-bang control with m switching points

$$0 \leq t_0^i \leq t_1^i \ldots \leq t_m^i \leq t_{m+1}^i = \ldots = t_n^i = T$$

and

$$u_i = (-1)^j \quad \text{for} \quad t \in [t_j^i, \; t_{j+1}^i) \quad j = 0,\ldots,m .$$

Step 2: Compute $\bar{u}_i$ which has k switching points

$$0 \leq s_0^i \leq s_1^i \ldots \leq s_k^i \leq s_{k+1}^i = \ldots = s_n^i = T$$

(if $k > n$ substitute n by $n = k + 1$ and complete the vector of switching points of $u_i$ by $t_j^i = T$ for $j = n + 1, \ldots, k + 1$) and

$$\bar{u}_i(t) = (-1)^j \quad \text{for} \quad t \in [s_j^i, s_{j+1}^i], \quad j = 0,\ldots,k .$$

Step 3: Define for $\alpha \in [0,1]$

$$t_j^i(\alpha) = \alpha t_j^i + (1 - \alpha) t_j^i , \quad j = 0,\ldots,n$$

and

$$u(\alpha)(t) = (-1)^j \quad \text{for} \quad t \in [t_j^i(\alpha), t_{j+1}^i(\alpha)), \quad j = 0,\ldots,n .$$

Compute $\alpha_i \in [0,1]$ such that

$$F(u(\alpha_i)) \leq F(u(\alpha)) \qquad \text{for all } \alpha \in [0,1] .$$

Step 4: Set

$$u_{i+1} = u(\alpha_i) ,$$

increase i by 1 and go to step 1.

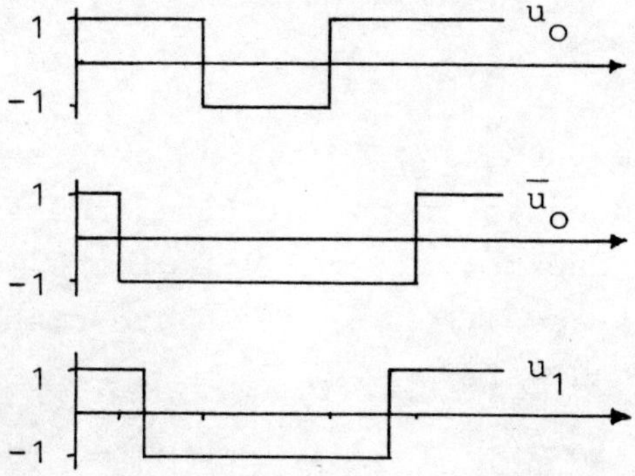


Figure 7.2: Variation of switching points

Although we do not prove convergence, the method has good global behavior in determining the shape of an optimal control and the number of switching points. When the gradient $g_i$ has finitely many switches, the method has the advantage that a bang-bang control is obtained at each iteration.

If the approximate behavior of the optimal control is determined, i.e. if $u_i$ and $\bar{u}_i$ for example have the same number of sign changes and both begin with the same value +1 or -1, the switching points can be introduced as variables $t_1, \ldots, t_m$. The optimality criterion for convex F yields that $\hat{u}$ is optimal if and only if

$$\langle \nabla F(\hat{u}), u - \hat{u} \rangle \geq 0 \qquad \text{for all } u \in U$$

or

$$\hat{u}(t) = - \operatorname{sgn} \nabla F(\hat{u})(t) .$$

If $\hat{u}$ has m switches, then $\hat{u}$ is optimal if and only if

$$\nabla F(u(\hat{t}_1,\ldots,\hat{t}_m))(\hat{t}_i) = 0 \qquad i = 1,\ldots,m \tag{7.37}$$

and the sign is correct in $[0,\hat{t}_0)$. Here $u(\hat{t}_1,\ldots,\hat{t}_m)$ denotes the function which is $\varepsilon(-1)^i$ in $[\hat{t}_i, \hat{t}_{i+1})$ for $i = 0,\ldots, m-1$, $\hat{t}_0=0$, $\varepsilon = +1$ or $\varepsilon = -1$.

(7.37) is a nonlinear system of equations for which Newton's method is a potential solution technique.

## 7.5 COMMENTS

Section 7.1: The Franke-Wolfe method or conditional gradient method was developed in [1] and extended to infinite dimensional spaces in 1965 by Demyanov and Rubinov [2] and Valadier [3]. Further variants of this method and its application to control problems were also treated in [4] - [6].

Section 7.2: The step length rule (7.23) is investigated in [7].

Section 7.3: For an example that the linear convergence rate cannot be improved in general, cf. [8]. An interesting connection between the smoothness of the optimal control and the convergence rate is proved recently in [9].

Section 7.4: The method in Algorithm 7B is presented in [10] and successfully applied to linear parabolic control problems. A related method is proposed in [11] and the convergence proof appeared in [12].

# 8 Projection methods

## 8.1 MOTIVATION

As in the preceding chapter we treat the following constrained optimization problem:

Let U be a nonempty, closed, convex subset of a Hilbert space H and F an objective functional

$$F : U \to \mathbb{R} .$$

Find $\hat{u} \in U$ such that

$$F(\hat{u}) \leq F(u) \qquad \text{for all } u \in U .$$

The approach we shall pursue in this chapter is to project vectors of H into the feasible set U. The projection is an operator P,

$$P : H \to U ,$$

where for each $v \in H$ the image Pv is the unique element of U such that

$$\|Pv - v\| \leq \|u - v\| \qquad \text{for all } u \in U , \tag{8.1}$$

i.e. Pv is the vector in U which has the smallest distance to v.

The projection is characterized by the following inequality

$$\langle Pv - v, u - Pv \rangle \geq 0 \qquad \text{for all } u \in U . \tag{8.2}$$

With respect to optimality we have the following theorem which yields an optimality test for algorithms:

<u>THEOREM 8.1:</u> Let $F : H \to H$ be Fréchet differentiable. If $\hat{u} \in U$ is is optimal, then for each $\alpha \in (0,\infty)$

$$P(\hat{u} - \alpha \nabla F(\hat{u})) = \hat{u} . \tag{8.3}$$

If F is convex on U, and (8.3) holds for some $\alpha > 0$, then $\hat{u}$ is optimal.

Proof: The condition

$$\langle \nabla F(\hat{u}), u - \hat{u} \rangle \geq 0 \qquad \text{for all } u \in U$$

is necessary for optimality of $\hat{u}$ and sufficient, if F is convex. This can be rewritten for any $\alpha \in \mathbb{R}$ as

$$\langle \hat{u} - (\hat{u} - \alpha \nabla F(\hat{u})), u - \hat{u} \rangle \geq 0 \qquad \text{for all } u \in U$$

which is equivalent to

$$\hat{u} = P(\hat{u} - \alpha \nabla F(\hat{u}))$$

using (8.2).

The straightforward generalization of descent methods for unconstrained problems is to find for a given descent direction $v_i$,

$$\langle \nabla F(u_i), v_i \rangle < 0 ,$$

a step length such that the value at the projected vector $u_i + \alpha_i v_i$ is minimal,

$$F(P(u_i + \alpha_i v_i)) \leq F(P(u_i + \alpha v_i))$$

for all $\alpha > 0$.

Hence, the basic form of a gradient projection method is the following:

Algorithm 8A:

Step 0: Select $u_0 \in U$, $i = 0$.

Step 1: Set

$$v_i = - \nabla F(u_i).$$

Step 2: Compute $\alpha_i$ by some step length rule.

Step 3: Set

$$u_{i+1} = P(u_i + \alpha_i v_i).$$

Step 4: Set $i = i + 1$ and go to 1.

By Theorem 8.1, the algorithm terminates if

$$u_{i+1} = u_i . \qquad (8.4)$$

The property, that at each iteration an actual descent of the value of the objective is achieved, has to be proved in connection with the step length rule that is used. A minor disadvantage of Algorithm 8A lies in step 2, where each evaluation of the function value requires a projection. It is possible to overcome this difficulty by projected gradient methods:

Algorithm 8B:

Step 0: Select $u_0 \in U$, $\lambda > 0$, $i = 0$.

Step 1: Set

$$\bar{u}_i = P(u_i - \lambda \nabla F(u_i)) ,$$

$$v_i = \bar{u}_i - u_i .$$

Step 2: Determine $\alpha_i \in (0,1]$ by some step length rule.

Step 3: Set

$$u_{i+1} = u_i + \alpha_i v_i .$$

Step 4: Increase i by 1 and go to step 1.

The main advantage is that we have a descent direction $v_i$ for any $\lambda$.

LEMMA 8.2: Let F be Fréchet differentiable on U. If $u_i \in U$ and

$$v_i = P(u_i - \lambda \nabla F(u_i)) - u_i \neq \Theta \tag{8.5}$$

for some $\lambda > 0$, then

$$<\nabla F(u_i), v_i> \leq - \frac{1}{\lambda} \|v_i\|^2 < 0 . \tag{8.6}$$

Proof: (8.2) implies with

$$\bar{u}_i = P(u_i - \lambda \nabla F(u_i))$$

$$0 \leq <\bar{u}_i - u_i + \lambda \nabla F(u_i), u_i - \bar{u}_i>$$

$$= - \|\bar{u}_i - u_i\|^2 - \lambda <\nabla F(u_i), v_i>$$

and

$$<\nabla F(u_i), v_i> \leq - \lambda^{-1} \|v_i\|^2 ,$$

showing the descent property.

## 8.2 GLOBAL CONVERGENCE

Let us first consider Algorithm 8A. We have to develop a convergence criterion for the sequence $\{u_i\}_{\mathbb{N}}$ that yields convergence of the function values to the minimal value.

LEMMA 8.3: Let F be convex and Fréchet differentiable on U, U bounded, $\{u_i\}_{\mathbb{N}} \subseteq U$ generated by Algorithm 8A and with step sizes $\alpha_i$ fulfilling

$$\alpha_i \geq \alpha^* > 0 \qquad \text{for all } i \in \mathbb{N} \tag{8.7}$$

and some $\alpha^* \in \mathbb{R}$. If

$$\lim_{i \to \infty} \|u_{i+1} - u_i\| = 0 , \tag{8.8}$$

then

$$\lim_{i \to \infty} F(u_i) = \inf_{u \in U} F(u) . \tag{8.9}$$

Proof: By the convexity of F, for each $u \in U$

$$\begin{aligned} F(u_i) - F(u) &\leq \langle \nabla F(u_i), u_i - u \rangle \\ &= \langle \nabla F(u_i), u_i - u_{i+1} \rangle + \frac{1}{\alpha_i} \langle u_i - \alpha_i \nabla F(u_i) - u_{i+1}, u - u_{i+1} \rangle \\ &\quad - \frac{1}{\alpha_i} \langle u_i - u_{i+1}, u - u_{i+1} \rangle \\ &\leq \langle \nabla F(u_i), u_i - u_{i+1} \rangle - \frac{1}{\alpha_i} \langle u_i - u_{i+1}, u - u_{i+1} \rangle , \end{aligned}$$

using step 3 of Algorithm 8A with (8.2). Thus,

$$F(u_i) - F(u) \leq \langle u_i - u_{i+1}, \nabla F(u_i) - \frac{1}{\alpha_i} (u - u_{i+1}) \rangle$$

and taking the supremum on both sides

$$0 \leq F(u_i) - \inf_{u \in U} F(u) \leq \|u_i - u_{i+1}\| \left( \|\nabla F(u_i)\| + \frac{2\nu}{\alpha_i} \right)$$

$$\nu = \sup_{u \in U} \|u\| . \tag{8.10}$$

If (8.8) holds, then inequality (8.10) immediately implies (8.9).

Lemma 8.3 indicates that we have to verify condition (8.8).

THEOREM 8.4: Let F be convex and Fréchet differentiable on U, with a Lipschitz-condition

$$\|\nabla F(u) - \nabla F(v)\| \le k \|u - v\| \qquad \text{for all } u, v \in U ,$$

U bounded, $\{u_i\}_{\mathbb{N}} \subset U$ generated by Algorithm 8A and with step-size $\alpha_i$ fulfilling for some small $\varepsilon_1, \varepsilon_2 > 0$

$$\frac{2}{k} - \varepsilon_1 \ge \alpha_i \ge \varepsilon_2 > 0 \qquad \text{for all } i \in \mathbb{N} . \tag{8.11}$$

Then

$$F(u_{i+1}) < F(u_i) \qquad \text{for all } i \in \mathbb{N}$$

and

$$\lim_{i \to \infty} F(u_i) = \inf_{u \in U} F(u) .$$

Proof: From Taylor expansion we have

$$F(u_{i+1}) - F(u_i) \le \langle \nabla F(u_i), u_{i+1} - u_i \rangle + \frac{k}{2} \|u_{i+1} - u_i\|^2$$

$$= - \frac{1}{\alpha_i} \langle u_i - \alpha_i \nabla F(u_i) - u_{i+1}, u_{i+1} - u_i \rangle$$

$$+ \left( \frac{k}{2} - \frac{1}{\alpha_i} \right) \|u_{i+1} - u_i\|^2 \le \left( \frac{k}{2} - \frac{1}{\alpha_i} \right) \|u_{i+1} - u_i\|^2 ,$$

using step 3 of Algorithm 8A and (8.2). Using (8.11) yields

$$F(u_{i+1}) - F(u_i) \le - \frac{\varepsilon_1}{2} \left( \frac{2}{k} - \varepsilon_1 \right)^{-1} \|u_{i+1} - u_i\|^2 < 0 . \tag{8.12}$$

Therefore, $\{F(u_i)\}_{\mathbb{N}}$ is a decreasing sequence, hence convergent and from (8.12)

$$\lim_{i \to \infty} \|u_{i+1} - u_i\|^2 = 0 .$$

Then Lemma 8.3 yields the desired statement.

Let us turn our attention to Algorithm 8B.

LEMMA 8.5: Let F be convex and Fréchet differentiable on U, U bounded, $\{u_i\}_{\mathbb{N}} \subseteq U$ generated by Algorithm 8B. If

$$\lim_{i \to \infty} \|\bar{u}_i - u_i\| = \lim_{i \to \infty} \|v_i\| = 0 , \tag{8.13}$$

then

$$\lim_{i \to \infty} F(u_i) = \inf_{u \in U} F(u) .$$

Proof: For each $u \in U$, using convexity,

$$F(u_i) - F(u) \le \langle \nabla F(u_i), u_i - u \rangle$$

$$= \langle \nabla F(u_i), u_i - \bar{u}_i \rangle + \frac{1}{\lambda} \langle u_i - \bar{u}_i - \lambda \nabla F(u_i), u - \bar{u}_i \rangle$$

$$- \frac{1}{\lambda} \langle u_i - \bar{u}_i, u - \bar{u}_i \rangle$$

$$\le \langle \nabla F(u_i), u_i - \bar{u}_i \rangle - \frac{1}{\lambda} \langle u_i - \bar{u}_i, u - \bar{u}_i \rangle ,$$

by step 2 of Algorithm 8B and (8.2).

Taking the supremum on both sides we have

$$0 \le F(u_i) - \inf_{u \in U} F(u) \le \langle \nabla F(u_i), v_i \rangle + \frac{2}{\lambda} \|v_i\| \nu ,$$

$$\nu = \sup_{u \in U} \|u\| .$$

Then assumption (8.13) yields the desired equation.

From this lemma it is easy to recognize the importance of condition (8.6)

$$\langle \nabla F(u_i), v_i \rangle \le - \frac{1}{\lambda} \|v_i\|^2 \le 0. \tag{8.6}$$

If we can assure, that

$$\lim_{i \to \infty} \langle \nabla F(u_i), v_i \rangle = 0 , \tag{8.14}$$

then (8.6) implies that assumption (8.13) of Lemma 8.5 holds. Recall that

$$v_i = \bar{u}_i - u_i ,$$

where $\bar{u}_i \in U$, and that $v_i$ is a descent direction. This is a similar situation as in the conditional gradient method of Chapter 7. If we use one of the step size rules of Section 7.2, except that from Lemma 7.5, then we obtain that (8.14) holds.

THEOREM 8.6: Let F be convex and Fréchet differentiable on U, U bounded, $\nabla F(u)$ Lipschitz continuous on U, $\{u_i\}_{\mathbb{N}} \subseteq U$ generated by Algorithm 8B with one of the following step length rules:

a) "exact" step length rule (7.10),

b) Goldstein's step length rule (7.12) - (7.13),

c) Powell's step length rule (7.14) - (7.16).

Then

$$\lim_{i \to \infty} F(u_i) = F(\hat{u}) .$$

Proof: As shown in the proofs of Lemma 7.2 - 7.4 we have

$$\lim_{i \to \infty} \langle \nabla F(u_i), v_i \rangle = 0 .$$

With Lemma 8.5 and (8.6) we obtain the statement to be proven.

## 8.3 CONVERGENCE RATES

As in Chapter 7 the methods presented here are derived using first order expansion of the cost functional. Hence it is reasonable to expect at most a linear convergence rate.

THEOREM 8.7: Let F be twice Fréchet differentiable on U and satisfy for some m, M > 0

$$m \langle v,v \rangle \le \langle v, \nabla^2 F(u)v \rangle \le M \langle v,v \rangle$$

for all $u \in U$, $v \in H$. If $\{u_i\} \subseteq U$ is a sequence of vectors generated by Algorithm 8A such that for some $\varepsilon_1, \varepsilon_2 > 0$

$$\frac{2}{M} - \varepsilon_1 \ge \alpha_i \ge \varepsilon_2 > 0 \quad \text{for all } i \in \mathbb{N} , \tag{8.15}$$

then $\{u_i\}$ converges at a linear rate to the optimal point $\hat{u}$.

Proof: The relation

$$u_{i+1} = P(u_i - \alpha_i \nabla F(u_i))$$

and (8.2) yield

$$\|u_{i+1} - \hat{u}\|^2 = \langle u_i - \alpha_i \nabla F(u_i) - u_{i+1}, \hat{u} - u_{i+1} \rangle$$

$$+ \langle \alpha_i \nabla F(u_i) + \hat{u} - u_i, \hat{u} - u_{i+1} \rangle$$

$$\le \langle \alpha_i \nabla F(u_i) + \hat{u} - u_i, \hat{u} - u_{i+1} \rangle .$$

Since $\hat{u}$ is optimal,

$$\langle \nabla F(\hat{u}), \hat{u} - u_{i+1} \rangle \le 0$$

and

$$\|u_{i+1} - \hat{u}\|^2 \le \langle \hat{u} - u_i + \alpha_i(\nabla F(u_i) - \nabla F(\hat{u})), \hat{u} - u_{i+1}\rangle .$$

Therefore

$$\|u_{i+1} - \hat{u}\| \le \|\hat{u} - u_i + \alpha_i(\nabla F(u_i) - \nabla F(\hat{u}))\|$$

$$\le \sup_{0 \le \lambda \le 1} \|I - \alpha_i \nabla^2 F(\hat{u} + \lambda(u_i - \hat{u}))\| \; \|u_i - \hat{u}\| , \tag{8.16}$$

by the mean value theorem. But

$$\sup_{0 \le \lambda \le 1} \max_{\|v\| = 1} |\langle v,v\rangle - \alpha_i \langle v, \nabla^2 F(\hat{u} + \lambda(u_i - \hat{u})v\rangle|$$

$$\le \max \{|1 - \alpha_i m|, |1 - \alpha_i M|\} = \rho. \tag{8.17}$$

It remains to show that $\rho$ in (8.17) is smaller than 1. This holds if and only if (8.15) is true

$$|1 - \alpha_i M| < 1 \iff 0 < \alpha_i < \frac{2}{M} . \tag{8.18}$$

Since (8.18) has to hold strictly for all $i \in \mathbb{N}$, (8.15) is the appropriate condition. (8.16) and (8.17) yield that there is $\sigma \in (0,1)$ such that

$$\|u_{i+1} - \hat{u}\| \le \sigma \|u_i - \hat{u}\| \qquad \text{for all } i \in \mathbb{N} .$$

In the general case, we prove a slower rate of convergence of the function values.

THEOREM 8.8: Let F be convex and Fréchet differentiable with Lipschitz continuous derivative on U. If $\{u_i\}_{\mathbb{N}}$ is a sequence generated by 8B with a step length rule by Goldstein, Powell or (7.18), then there is a constant $\zeta > 0$ and an index $i_0 \in \mathbb{N}$ such that for $i \ge i_0$

$$|F(u_i) - \inf_{u \in U} F(u)| \le \frac{\zeta}{i} .$$

Proof: From the convexity we obtain for $u \in U$

$$F(u_i) - F(u) \le \langle \nabla F(u_i), u_i - u\rangle$$

$$= \langle \nabla F(u_i), u_i - \bar{u}_i\rangle + \frac{1}{\lambda} \langle u_i - \lambda \nabla F(u_i) - \bar{u}_i, u - \bar{u}_i\rangle$$

$$- \frac{1}{\lambda} \langle u_i - \bar{u}_i, u - \bar{u}_i \rangle$$

$$\leq \langle \nabla F(u_i), u_i - \bar{u}_i \rangle - \frac{1}{\lambda} \langle u_i - \bar{u}_i, u - \bar{u}_i \rangle \tag{8.19}$$

using

$$\bar{u}_i = P(u_i - \lambda \nabla F(u_i))$$

and (8.2). Hence with

$$\rho_i = F(u_i) - \inf_{u \in U} F(u)$$

(8.19) implies

$$\rho_i \leq - \langle \nabla F(u_i), v_i \rangle + \frac{1}{\lambda} \sup_{u \in U} \langle v_i, u - \bar{u}_i \rangle$$

and with

$$\nu = \sup_{u \in U} \|u\| ,$$

$$\rho_i \leq \gamma \|v_i\| \tag{8.20}$$

for some constant $\gamma$.

The step size rules yield, that for some $\sigma > 0$, $i_o \in \mathbb{N}$ (see Lemma 7.3 and 7.4)

$$F(u_{i+1}) - F(u_i) \leq \sigma \langle \nabla F(u_i), v_i \rangle$$

or

$$\rho_{i+1} - \rho_i \leq \sigma \langle \nabla F(u_i), v_i \rangle \qquad \text{for all } i \geq i_o . \tag{8.21}$$

Hence (8.20), (8.6) and (8.21) imply for $i \geq i_o$

$$\rho_i^2 \leq \gamma^2 \|v_i\|^2 \leq - \lambda \gamma^2 \langle \nabla F(u_i), v_i \rangle \leq \lambda \gamma^2 \sigma (\rho_i - \rho_{i+1}).$$

This is a similar inequality as (7.33) from which we deduce that there exists $\zeta \in \mathbb{R}$ such that for $i \geq i_o$

$$\rho_i \leq \frac{\zeta}{i} .$$

## 8.4 COMMENTS

Sections 8.1 - 8.3: These methods are also known as Uzawa's and Rosen's methods [1], [2]. Since projection methods and their numerous variants are well known, we shall confine our references for infinite dimensional problems to the books of Daniel [3] and Demyanov and Rubinov [4]. For a further discussion of convergence see Levitin and Polyak [5].

# 9 Approximation – type methods

## 9.1 NEWTON'S METHOD

This chapter treats optimal control problems that are formulated as approximation problems. The general form of these problems is as follows:

Given a nonlinear map $G : W \to E$, where $W \subseteq \mathbb{R}^n$ is non-empty and E is a normed linear space:

Find $\hat{w} \in W$ such that $(P_A)$

$$\|G(\hat{w})\| \leq \|G(w)\| \qquad \text{for all } w \in W .$$

In order to solve this problem numerically, we begin with an iterate vector $w_i \in W$ and try to find a better approximation $w_{i+1} \in W$. In case that the norm in $(P_A)$ is Fréchet differentiable, we define $F(w) = \|G(w)\|$ and can apply the previous algorithms, provided $W = \mathbb{R}^n$. If $\|\cdot\|$ is not Fréchet differentiable, e.g. max- or $L_1$-norm, then a Newton-like method would linearize G at $w_i$ and solve the resulting linear approximation problem. We suppose that G is Fréchet differentiable on some open set containing W.

We assume W is a convex set.

Step 0. Select $w_0 \in W$, set $i = 0$.

Step 1. Find $w_{i+1} \in W$ such that for all $w \in D$

$$\|G(w_i) + G'_{w_i}(w_{i+1} - w_i)\| \leq \|G(w_i) + G'_{w_i}(w - w_i)\|.$$

Step 2. Set $i = i + 1$, go to 1.

In the practical implementation we usually insert a line search between steps 1 and 2:

Find $\alpha_{i+1} \in [0,1]$, such that for all $\alpha \in [0,1]$

$$\|G(w_i + \alpha_{i+1}(w_{i+1} - w_i))\| \leq \|G(w_i + \alpha(w_{i+1} - w_i))\| .$$

Then $w_i + \alpha_{i+1}(w_{i+1} - w_i)$ replaces $w_{i+1}$ during the next iteration. For many of the control approximation problems the calculation of the derivative $G'_{w_i}$ of G at each point can be rather time consuming. In particular, if the operator G is defined via a nonlinear differential equation, the determination of the derivative requires the solution of a system of linear differential equations. Therefore, it is desirable to alter the algorithm in such a way that $G'_{w_i}$ is replaced by some approximation, e.g. a variable metric-like update. Since in general $G'_w$ does not have any symmetry properties, we can use only requirements such as small dimension of the range of the correction operator and the quasi-Newton equation for the determination of the update. Hence, the prototype of a variable metric algorithm for constrained optimization follows.

ALGORITHM 9 A:

Step 0. Select $w_o \in W$, $B_o : \mathbb{R}^n \to E$ linear, bounded, $i = 0$.

Step 1. Find $w_{i+1} \in W$ such that for all $w \in W$

$$\|G(w_i) + B_i(w_{i+1} - w_i)\| \leq \|G(w_i) + B_i(w - w_i)\| .$$

Step 2. Compute $B_{i+1}$ by a variable metric update formula.

Step 3. Set $i = i + 1$, go to 1.

The quasi-Newton equation follows from linearization of G at $w_i$:

$$G(w_{i+1}) \cong G(w_i) + G'_{w_i}(w_{i+1} - w_i) .$$

Hence with

$$p_i = w_{i+1} - w_i , \quad y_i = G(w_{i+1}) - G(w_i) \tag{9.1}$$

and $B_{i+1}$ to be an approximation of $G'_{w_i}$

$$y_i = B_{i+1} p_i . \tag{9.2}$$

Since symmetry or positive definiteness properties of $G'_w$ are not generally available, in contrast to the unconstrained case, we confine our treatment to rank-one corrections for the updates.

LEMMA 9.1: Each $e \in E$, $a \in \mathbb{R}^n$ defines a linear continuous operator $]e,a[ \; : \mathbb{R}^n \to E$ by

$$]e,a[\; w = \langle a,w\rangle \, e \qquad \text{for all } w \in \mathbb{R}^n \;.$$

The analog of Lemma 3.2 is

LEMMA 9.2: Each linear continuous operator $T : \mathbb{R}^n \to E$ which satisfies $Tp = z$ for some $p \in \mathbb{R}^n$, $z \in E$ and dim range $T = 1$ is given by

$$T = \frac{]z,a[}{\langle a,p\rangle} \;, \quad a \in \mathbb{R}^n \;, \quad \langle a,p\rangle \neq 0 \;. \tag{9.3}$$

Since we shall use the fact that $G'_w$ is injective for the convergence proof, we discuss this property for $\bar{B}$.

Let $\bar{B}$ be given by

$$\bar{B} = \rho B + \sigma \frac{]z,a[}{\langle a,p\rangle} \;, \rho,\sigma \in \mathbb{R} \;, \quad a \in \mathbb{R}^n \;, \quad \langle a,p\rangle \neq 0 \;. \tag{9.4}$$

LEMMA 9.3: Let $B : \mathbb{R}^n \to E$ be injective, $z \neq \Theta$, $\rho \neq 0$.

Then $\bar{B}$ given by (9.4) is also injective if and only if

$$\sigma \langle a,B^{-1}z\rangle + \rho\langle a,p\rangle \neq 0 \;. \tag{9.5}$$

Proof: Assume for some $w \in \mathbb{R}^n$ that

$$\Theta = \bar{B}w = \rho Bw + \sigma \langle a,p\rangle^{-1} \langle a,w\rangle \, z \;.$$

If $Bw$ and $z$ are linearly independent, then $Bw = \Theta$ and by supposition $w = \Theta$. Otherwise, there is $\zeta \in \mathbb{R}$ such that $Bw = \zeta z$, $\zeta \neq 0$ and hence

$$\begin{aligned} \Theta &= (\zeta\rho + \sigma \langle a,p\rangle^{-1} \langle a,w\rangle) \, z \\ &= \zeta(\rho + \sigma \langle a,p\rangle^{-1} \langle a,B^{-1}z\rangle) \, z \;. \end{aligned}$$

This implies $w = \Theta$ if and only if (9.5) is satisfied .

REMARK: If $z = y - \rho Bp$, as usual, then (9.5) is equivalent to

$$\rho(1 - \sigma) + \sigma \frac{\langle a,B^{-1}p\rangle}{\langle a,p\rangle} \neq 0 \;.$$

Hence an appropriate adjustment of $\rho$ or $\sigma$ preserves the injectivity of the updates.

## 9.2 STRONG UNIQUENESS

Since we do not have differentiability of the norms we need an additional property which yields the estimates used for convergence proofs.

DEFINITION 9.4: A vector $\hat{w} \in W$ is a locally strongly unique best approximation if there are $\kappa, \varepsilon > 0$, such that for all $w \in W$, $\|w - \hat{w}\| \leq \varepsilon$

$$\|G(w)\| \leq \kappa ( \|G(w - \hat{w})\| - \|G(\hat{w})\| ) . \tag{9.6}$$

If (9.6) holds for all $w \in W$, then $\hat{w}$ is a globally strongly unique best approximation.

Assuming G to be Fréchet differentiable, the local uniqueness is characterized by a global property of the linearized problem, see e.g. [1].

LEMMA 9.5: If G is Fréchet differentiable at $\hat{w} \in W$, the vector $\hat{w}$ is a locally strongly unique best approximation of

$$\min_{w \in W} \|G(w)\|$$

if and only if $\hat{w}$ is a globally strongly unique best approximation of

$$\min_{w \in W} \|G(\hat{w}) + G'_{\hat{w}} (w - \hat{w})\| ,$$

i.e. there is a $\kappa > 0$ such that for all $w \in W$

$$\|G'_{\hat{w}} (w - \hat{w})\| \leq \kappa ( \|G(\hat{w}) + G'_{\hat{w}} (w - \hat{w})\| - \|G(\hat{w})\| ). \tag{9.7}$$

In approximation theory there are several sufficient conditions for strong uniqueness. We are mainly interested in the case of $E = C(I)$, I some closed and bounded interval, equipped with the maximum-norm. The most common condition for strong uniqueness as in (9.7) is the Haar condition.

LEMMA 9.6: $\{g_1, \ldots, g_n\} \subset C(I)$, I compact interval, is a Haar system if and only if

$$\sum_{i=1}^{n} \alpha_i g_i \ , \qquad \alpha \in \mathbb{R}^n$$

vanishes identically in I or has at most n - 1 zeroes in I. If $\{G'_{\hat{w}}(e_i)\}_{i=1}^n$, $e_i$ linearly independent, is a Haar system, then strong uniqueness i.e. (9.7) holds.

A proof of these lemma can be found in [2]. In our application to parabolic control problems we will give a further discussion of this property.

For the algorithms we obtain the following estimate:

<u>LEMMA 9.7:</u> Let $w = w_i \in W$ and $\overline{w} = w_{i+1}$ be constructed by step 1 of algorithm 9A. If $\hat{w}$ is a locally strongly unique best approximation, then with $B = B_i$ for some $\kappa > 0$ (independent of i)

$$\|G'_{\hat{w}}(\overline{w} - \hat{w})\| \le \kappa( \|G(w) - G(\hat{w}) + B(\hat{w} - w)\| + \|G(\hat{w}) - G(w) + G'_{\hat{w}}(\overline{w} - \hat{w}) - B(\overline{w} - w)\| ) . \tag{9.8}$$

<u>Proof:</u> (9.7) yields

$$\|G'_{\hat{w}}(\overline{w} - \hat{w})\| \le \kappa( \|G(\hat{w}) + G'_{\hat{w}}(\overline{w} - \hat{w})\| - \|G(\hat{w})\| )$$

$$\le \kappa( \|G(w) + B(\overline{w} - w)\| - \|G(\hat{w})\| + \|G(\hat{w}) - G(w) + G'_{\hat{w}}(\overline{w} - \hat{w}) - B(\overline{w} - w)\| )$$

and by step 1 of algorithm 9A

$$\le \kappa( \|G(w) + B(\hat{w} - w)\| - \|G(\hat{w})\| + \|G(\hat{w}) - G(w) + G'_{\hat{w}}(\overline{w} - \hat{w}) - B(\overline{w} - w)\| ) ,$$

from which (9.8) follows.

## 9.3 QUADRATIC CONVERGENCE RATE

The various convergence rates depend upon how well $B_i$ approximates the derivative $G'_{\hat{w}}$ at the optimal point.

Sometimes we assume that G has a Lipschitz continuous derivative in a neighborhood of $\hat{w}$, i.e. there is a neighborhood N of $\hat{w}$ and some $\mu > 0$ with

$$\|G'_{\hat{w}} - G'_e\| \le \mu \; \|\hat{w} - e\| \qquad \text{for all } e \in N . \tag{9.9}$$

THEOREM 9.8: Let the condition (9.9) on G hold and let $G'_{\hat{w}}$ be injective and $\{w_i\}_{\mathbb{N}} \subseteq W$ constructed by Algorithm 9A such that $w_i$ converges to a locally strongly unique best approximation $\hat{w}$. If for some $\nu > 0$

$$\|B_i - G'_{\hat{w}}\| \le \nu \; \|w_i - \hat{w}\| \qquad \text{for all } w \in W . \tag{9.10}$$

then the convergence rate of $w_i$ is quadratic.

Proof: From Lemma 9.7 and (9.8) for i large enough

$$\|G'_{\hat{w}}\,(w_{i+1} - \hat{w})\| - \kappa \; \|(G'_{\hat{w}} - B_i)(w_{i+1} - \hat{w})\|$$

$$\le 2\kappa \; \|G(\hat{w}) - G(w_i) - B_i\,(\hat{w} - w_i)\|$$

$$\le 2\kappa \; \left( \|G(\hat{w}) - G(w_i) - G'_{\hat{w}}\,(\hat{w} - w_i)\| + \|(G'_{\hat{w}} - B_i)(\hat{w} - w_i)\| \right)$$

and for some $\bar{w}_i \in [w_i, \hat{w}] = \{w \in W \mid w = w_i + \lambda\,(\hat{w} - w_i),\ \lambda \in [0,1]\}$

$$\le 2\kappa \; \left( \|(G'_{\bar{w}_i} - G'_{\hat{w}})(\hat{w} - w_i)\| + \|(G'_{\hat{w}} - B_i)(\hat{w} - w_i)\| \right)$$

with (9.9) and (9.10), for some $\kappa_1 > 0$

$$\le 2\kappa_1 \; \|w_i - \hat{w}\|^2 . \tag{9.11}$$

Since $G'_{\hat{w}}$ is injective, there is $\gamma > 0$ such that

$$\|G'_{\hat{w}}\,(w)\| \ge \gamma \; \|w\| \qquad \text{for all } w \in W . \tag{9.12}$$

From the convergence of $w_i$ and (9.10) we obtain that for i large enough

$$\kappa \, \|G'_{\hat{w}} - B_i\| \le \frac{\gamma}{2}$$

and hence from (9.11) and (9.12)

$$\|w_{i+1} - \hat{w}\| \le 4\kappa_1 \, \gamma^{-1} \; \|w_i - \hat{w}\|^2 .$$

This theorem shows that quadratic convergence is achieved in the case of the Newton version

$$B_i = G'_{w_i} ,$$

because (9.10) is true by the Lipschitz continuity of G', or if

$$B_i = G'_{\hat{w}} .$$

In the first case one has the disadvantage of calculating $G'_{w_i}$ at each iteration. The second case occurs in only rather exceptional problems where the derivative at the optimal point is known but not the optimum itself.

## 9.4 LINEAR CONVERGENCE RATE

We shall give some results on linear convergence rate before treating the case of superlinear convergence.

THEOREM 9.9: Let (9.9) be satisfied and $\{w_i\}_{\mathbb{N}} \in W$ be a sequence generated by Algorithm 9A where the update formula fulfills for some $\beta, \delta \in \mathbb{R}$

$$\|B_{i+1} - G_{\hat{w}}\| \leq (1 + \beta\lambda_i) \|B_i - G'_{\hat{w}}\| + \delta(1 + \|B_i\|)\lambda_i \quad (9.13)$$

with $\lambda_i = \max \{ \|w_{i+1} - \hat{w}\| , \|w_i - \hat{w}\| \}$ for all $i \in \mathbb{N}$.

If $\hat{w} \in W$ is a locally strongly unique best approximation and $G'_{\hat{w}}$ is injective, then $w_i$ converges to $\hat{w}$ linearly and for some $\alpha \in (0,1)$, $\nu \in (0, \frac{\gamma}{\kappa})$,

$$\|w_{i+1} - \hat{w}\| \leq \alpha \|w_i - \hat{w}\| \qquad \text{for all } i \in \mathbb{N} \qquad (9.14)$$

and

$$\|B_i - G'_{\hat{w}}\| \leq \nu \qquad \text{for all } i \in \mathbb{N} , \qquad (9.15)$$

provided $\|w_0 - \hat{w}\|$ and $\|B_0 - G'_{\hat{w}}\|$ are small enough.

Proof: From Lemma 9.7 and (9.8) we obtain the line preceding (9.11), i.e. with (9.9) and (9.12)

$$(\gamma - \kappa \|G'_{\hat{w}} - B_i\|) \|w_{i+1} - \hat{w}\|$$

$$\leq 2\kappa (\mu \|w_i - \hat{w}\| + \|G'_{\hat{w}} - B_i\|) \|w_i - \hat{w}\| . \qquad (9.16)$$

Let

$$\|G'_{\hat{w}} - B_o\| \le \varepsilon \quad \text{and} \quad \|w_o - \hat{w}\| \le \varepsilon ,$$

with $\varepsilon$ so small such that the estimates in (9.9) hold. Then (9.16) gives

$$(\gamma - \kappa\varepsilon) \ \|w_1 - \hat{w}\| \le 2\kappa\varepsilon \ (1 + \mu) \ \|w_o - \hat{w}\| .$$

and for $\varepsilon \in (0, \frac{\gamma}{\kappa})$,

$$\|w_1 - \hat{w}\| \le \frac{2\kappa\varepsilon}{\gamma - \kappa\varepsilon} (1 + \mu) \ \|w_o - \hat{w}\| .$$

For given $\alpha \in (0,1)$ there is $\varepsilon > 0$ such that

$$\frac{2\kappa\varepsilon}{\gamma - \kappa\varepsilon} \ (1 + \mu) \le \alpha ,$$

and (9.14) holds for $i = 0$.

In order to fulfill (9.15) for $i = 0$ we only need to choose $\varepsilon \in (0, \nu)$.

Let us assume that (9.14) and (9.15) are true for all $i = 0,1,\dots,r-1$, where $r \in \mathbb{N}$.

Then by assumption for $i = 0,1,\dots,r-1$

$$\|w_{i+1} - \hat{w}\| \le \alpha \|w_i - \hat{w}\| \le \alpha^{i+1} \|w_o - \hat{w}\| \le \alpha^{i+1}\varepsilon, \qquad (9.17)$$

From (9.13)

$$\|B_r - G'_{\hat{w}}\| \le (1 + \beta\lambda_{r-1}) \|B_{r-1} - G'_{\hat{w}}\| + \delta(1 + \|B_{r-1}\|)\lambda_{r-1}$$

with (9.17) for $i = r-1$, $\lambda_{r-1} = \|w_{r-1} - \hat{w}\| \le \alpha^{r-1}\varepsilon$ ,

$$\|B_r - G'_{\hat{w}}\| \le \|B_{r-1} - G'_{\hat{w}}\| + \beta\alpha^{r-1}\varepsilon\nu + \delta(1 + \nu + \|G'_{\hat{w}}\|)\alpha^{r-1}\varepsilon$$

using this inequality iteratively with $\overline{\delta} = \delta(1 + \nu + \|G'_{\hat{w}}\|)$

$$\|B_r - G'_{\hat{w}}\| \le \|B_o - G'_{\hat{w}}\| + (\beta\varepsilon\nu + \overline{\delta}\varepsilon) \sum_{i=1}^{r} \alpha^{i-1}$$

and

$$\|B_r - G'_{\hat{w}}\| \le \varepsilon + \varepsilon \ (\beta\nu + \overline{\delta})(1 - \alpha)^{-1} . \qquad (9.18)$$

From this inequality it is clear that for $\varepsilon$ small enough and independent of $r$,

$$\|B_r - G'_{\hat{w}}\| \le \nu .$$

Therefore (9.16) implies with (9.18)

$$(\gamma - \kappa\nu) \|w_{r+1} - \hat{w}\| \le 2\kappa\ (\mu\alpha^r\varepsilon + \varepsilon + \varepsilon(\beta\nu + \bar{\delta})(1-\alpha)^{-1}) \|w_r - \hat{w}\| .$$

Moreover, if $\varepsilon$ is chosen from an interval, which is sufficiently small but independent of r, then (9.14) is true for i = r. Hence we have completed the proof by induction and (9.14) and (9.15) hold for all $i \in \mathbb{N}$ .

An application of the class of updates (9.4) will occur after the next section.

## 9.5 SUPERLINEAR CONVERGENCE

We recall superlinear convergence of $\{w_i\}_{\mathbb{N}}$ to $\hat{w}$ means

$$\lim_{i\to\infty} \frac{\|w_{i+1} - \hat{w}\|}{\|w_i - \hat{w}\|} = 0 .$$

In the next theorem we give a sufficient condition for superlinear convergence.

THEOREM 9.10: Let all the assumptions of Theorem 9.9 hold and the condition (9.13) on the convergence of $B_i$ to $G'_{\hat{w}}$ be replaced by

$$\lim_{i\to\infty} \frac{\|(B_i - G'_{\hat{w}})(w_{i+1} - w_i)\|}{\|w_{i+1} - w_i\|} = 0 . \qquad (9.19)$$

If we further assume that (9.14) and (9.15) hold, then $\{w_i\}_{\mathbb{N}}$ converges superlinearly to $\hat{w}$.

Proof: From Lemma 9.7 we obtain with (9.12)

$$\gamma\ \|w_{i+1} - \hat{w}\| \le \kappa\ (\|G(\hat{w}) - G(w_i) - B_i(\hat{w} - w_i)\| +$$

$$+ \|G(\hat{w}) - G(w_i) + G'_{\hat{w}}\ (w_{i+1} - \hat{w}) - B_i\ (w_{i+1} - w_i)\|)$$

$$\le \kappa\ (2\|G(\hat{w}) - G(w_i) - G'_{\hat{w}}\ (\hat{w} - w_i)\| +$$

$$+ 2\|(G'_{\hat{w}} - B_i)(w_{i+1} - w_i)\| + \|(G'_{\hat{w}} - B_i)(w_{i+1} - \hat{w})\|)$$

and with the mean value theorem and (9.9)

$$\le \kappa \; (2\mu \; \|\hat{w} - w_i\|^2 + 2\,\|(G'_{\hat{w}} - B_i)(w_{i+1} - w_i)\| +$$

$$+ \nu\,\|w_{i+1} - \hat{w}\|) \; .$$

Hence

$$(\gamma - \kappa\nu)\; \frac{\|w_{i+1} - \hat{w}\|}{\|w_{i+1} - w_i\|}$$

$$\le \kappa \; (2\mu \; \frac{\|\hat{w} - w_i\|^2}{\|w_{i+1} - w_i\|} + \frac{2\,\|(G'_{\hat{w}} - B_i)(w_{i+1} - w_i)\|}{\|w_{i+1} - w_i\|})$$

and with (9.14) $\quad \|w_{i+1} - w_i\| \ge (\frac{1}{\alpha} - 1)\,\|w_{i+1} - \hat{w}\|$

$$\le \kappa \; (2\mu \; \frac{\alpha}{1-\alpha} \; \frac{\|w_i - \hat{w}\|}{\|w_{i+1} - \hat{w}\|} \; \|w_i - \hat{w}\|$$

$$+ \frac{2\|(G'_{\hat{w}} - B_i)(w_{i+1} - w_i)\|}{\|w_{i+1} - w_i\|} \; ) \; .$$

If $w_i$ does not converge superlinearly to $\hat{w}$, i.e. there is a subsequence $\{w_i\}_\Lambda$ with $\Lambda \subset \mathbb{N}$ for which

$$\sigma_j = \|w_j - \hat{w}\| \;\; \|w_{j+1} - \hat{w}\|^{-1} \le \delta \qquad \text{for all } j \in \Lambda ,$$

then (9.19) and the estimate from above imply

$$0 = \lim_{j \in \Lambda} \frac{\|w_{j+1} - \hat{w}\|}{\|w_{j+1} - w_j\|} \ge \lim_{j \in \Lambda} \frac{1}{1+\sigma_j} \ge \frac{1}{1+\delta} \; ,$$

a contradiction. Hence, superlinear convergence is proven.

## 9.6 DISCUSSION OF UPDATES

All updates of the type (9.4) with a scaling factor are considered:

$$\overline{B} = \rho B + \sigma \frac{]z,a[}{\langle a,p\rangle} \; , \quad \sigma \in \mathbb{R} \, , \quad a \in \mathbb{R}^n \, , \; \langle a,p\rangle \neq 0 \; . \qquad (9.20)$$

In order to prove linear convergence we need to show that (9.13) holds for formulas of type (9.20). With $z = y - \rho Bp$ we derive from (9.20)

$$\overline{B} - G'_{\hat{w}} = (\rho B - G'_{\hat{w}})(I - \sigma \frac{]p,a[}{\langle a,p\rangle}) + \sigma \frac{]y - G'_{\hat{w}}\, p,a[}{\langle a,p\rangle}, \qquad (9.21)$$

where $I - \sigma \langle a,p\rangle^{-1}\, ]p,a[$ is defined as an operator on $\mathbb{R}^n$.

<u>LEMMA 9.11:</u> Let (9.9) be fulfilled, $\langle a,p\rangle \neq 0$. Then we have

$$\|\overline{B} - G'_{\hat{w}}\| \leq \|B - G'_{\hat{w}}\|\ \beta_1 + |1-\rho|\ \|B\|\ \beta_1 + \kappa\pi\lambda, \qquad (9.22)$$

where

$$\lambda = \max\{\|\overline{w} - \hat{w}\|, \|w - \hat{w}\|\},$$

$$\beta_1 = \max\{1, (1 - \sigma + \frac{1}{2}\sigma^2\pi^2 + [(1 - \sigma + \frac{\sigma^2}{4}\pi^2)\sigma^2\pi^2]^{1/2})^{1/2}\},$$

$$\pi = \|p\|\ \|a\|\ \langle a,p\rangle^{-1}.$$

<u>Proof:</u> From (9.21)

$$\|\overline{B} - G'_{\hat{w}}\| \leq \|\rho B - G'_{\hat{w}}\|\ \|I - \sigma \frac{]p,a[}{\langle p,a\rangle}\| + \sigma \frac{\|]y - G'_{\hat{w}}\, p,a[\|}{\langle p,a\rangle}, \qquad (9.23)$$

where the norm symbols denote the norms of the operators corresponding to their domain and range.

From the mean value theorem and (9.9)

$$\|]y - G'_{\hat{w}}\, p,a[\| \leq \|a\|\ \|G(\overline{w}) - G(w) - G'_{\hat{w}}(\overline{w} - w)\| \leq \kappa\|a\|\ \lambda\ \|p\|. \qquad (9.24)$$

We compute the norm of $I - \overline{\sigma}\,]p,a[$, $\overline{\sigma} = \sigma\langle p,a\rangle^{-1}$.

$$\phi = \sup \|x - \overline{\sigma}\langle a,x\rangle p\| \quad \text{for } \|x\| \leq 1. \qquad (9.25)$$

Hence for the optimal point $\hat{x}$ of (9.25), there is a $\mu \in \mathbb{R}$ such that

$$\hat{x} - \overline{\sigma}\langle a,\hat{x}\rangle p - \overline{\sigma}\langle p,\hat{x}\rangle a + \overline{\sigma}^2\langle p,p\rangle\langle a,\hat{x}\rangle a = \mu\hat{x}.$$

We have for $\mu > 0$, $\mu = \phi^2$ and from Lemma 3.8 $\mu = 1$ or

$$\mu = 1 - \bar{\sigma} \langle a,p\rangle + \frac{\bar{\sigma}^2}{2} \langle p,p\rangle \langle a,a\rangle$$

$$+ ((- \bar{\sigma} \langle p,p\rangle + \bar{\sigma}^2 \langle p,p\rangle \langle a,p\rangle)(- \bar{\sigma} \langle a,a\rangle)$$

$$+ \frac{1}{4} (\bar{\sigma}^2 \langle p,p\rangle \langle a,a\rangle)^2)^{1/2}$$

$$= 1 - \sigma + \frac{1}{2} \sigma^2 \pi^2 + [(1 - \sigma + \frac{\sigma^2}{4} \pi^2) \sigma^2 \pi^2]^{1/2} \quad . \qquad (9.26)$$

Furthermore,

$$(1 - \sigma + \frac{\sigma^2}{4} \pi^2) = (1 - \frac{\sigma}{2} \pi)^2 + (\pi - 1) \sigma$$

$$= (1 + \frac{\sigma}{2} \pi)^2 + (- \pi - 1) \sigma \geq 0 \ ,$$

whatever the sign of $\sigma$ is.

Since $\beta_1$ is a rather complicated expression, we shall ask under which condition $\beta_1$ equals 1. This holds if and only if

$$(1 - \sigma + \frac{\sigma^2}{4} \pi^2) \sigma^2 \pi^2 \leq (\sigma - \frac{1}{2} \sigma^2 \pi^2)^2$$

which is equivalent to

$$\sigma^2 \pi^2 \leq \sigma^2 \ .$$

However, $\pi^2 = 1$ is only true if vector a is a multiple of p. Since the length of a does not effect the update, a = p yields the version of the Broyden update in the class (9.21):

$$\bar{B} = \rho \ (B - \sigma \frac{]Bp,p[}{\langle p,p\rangle} + \sigma \frac{]y,p[}{\langle p,p\rangle}) \ . \qquad (9.27)$$

In the case $\sigma = 1$, $\beta_1$ also simplifies. Then $\beta_1 = \max \{1, \pi^2\} = \pi^2$.

<u>REMARK:</u> If $\{B_i\}_{\mathbb{N}}$ is updated by (9.27), and $\rho_i$ is such that for some $\eta$

$$|1 - \rho_i| \leq \eta \max \{ ||w_{i+1} - \hat{w}|| \ , \ ||w_i - \hat{w}|| \} \ , \qquad (9.28)$$

then (9.13) holds. Under the other assumptions of Theorem 9.9 we have linear convergence.

(9.28) is true, if we choose for example

$$\rho_i = 1 + \varepsilon_i , \qquad |\varepsilon_i| \le \eta \, \|p_i\| .$$

Having investigated the case of linear convergence we now consider superlinear convergence, which holds if condition (9.21) is true. For this latter purpose we need a better estimate than (9.22). The smallest possible value for $\beta_1$ is 1. However, in proofs treating finite dimensional optimization problems, the Frobenius-norm of $B - G'_{\hat{w}}$ is used and $\|\bar{B} - G'_{\hat{w}}\| - \|B - G'_{\hat{w}}\|$ is more sharply estimated. Since B maps $\mathbb{R}^n$ into a normal space, which in applications may not even be a Hilbert space, this procedure does not seem to apply. However, a useful estimate is obtained if we are working with linear functionals.

LEMMA 9.12: Let $\{B_i\}_{\mathbb{N}}$ be produced by (9.27) with $\rho_i = 1$, such that $\|B_i - G'_{\hat{w}}\|$ is bounded and

$$\sum_{i=1}^{\infty} \|w_i - \hat{w}\| < \infty , \quad \sigma_i \in (\varepsilon, 2 - \varepsilon) , \tag{9.29}$$

for some $\varepsilon \in (0,1)$.

Then for any linear continuous functional $\ell \in E^*$,

$$\lim_{i \to \infty} \frac{\ell((B_i - G'_{\hat{w}})p_i)}{\|p_i\|} = 0 , \tag{9.30}$$

i.e. $(B_i - G'_{\hat{w}})(\frac{p_i}{\|p_i\|})$ converges weakly to $\Theta$.

Proof: Select $\ell \in E^*$. Then (9.27) yields

$$\sup_{\|x\| \le 1} \ell((B_{i+1} - G'_{\hat{w}})x)$$

$$\le \sup_{\|x\| \le 1} \ell((B_i - G'_{\hat{w}})(x - \sigma_i \frac{\langle p_i , x\rangle}{\langle p_i , p_i\rangle} p_i)$$

$$+ \sup_{\|x\| \le 1} \sigma_i \, \ell(y_i - G'_{\hat{w}} \, p_i) \frac{\langle p_i , x\rangle}{\langle p_i , p_i\rangle} . \tag{9.31}$$

Since $C_i = B_i - G'_{\hat{w}} : \mathbb{R}^n \to E$, there are $c_j \in E$, $j = 1,\ldots,n$ such that $C_i w = \sum_{j=1}^{n} w_j c_j$ and

$$\ell(C_i w) = \langle \eta, w \rangle \quad , \quad \eta_i^{(j)} = \ell(c_j), \; j = 1,\ldots,n .$$

Therefore,

$$\sup_{\|x\| \le 1} \ell(C_i(x - \sigma_i \frac{\langle p_i, x\rangle}{\langle p_i, p_i\rangle} p_i))$$

$$= \sup_{\|x\| \le 1} \langle \eta_i, x - \sigma_i \frac{\langle p_i, x\rangle}{\langle p_i, p_i\rangle} p_i \rangle$$

$$= \|\eta_i - \sigma_i \frac{\langle p_i, \eta_i\rangle}{\langle p_i, p_i\rangle} p_i\|$$

$$= (\langle \eta_i, \eta_i\rangle - \sigma_i (\sigma_i - 2) \frac{\langle p_i, \eta_i\rangle^2}{\langle p_i, p_i\rangle})^{1/2} . \tag{9.32}$$

The last term in (9.31) can be estimated as follows

$$\sup_{\|x\| \le 1} \sigma_i \, \ell(y_i - G'_{\hat{w}} p_i) \frac{\langle p_i, x\rangle}{\langle p_i, p_i\rangle}$$

$$\le \sigma_i \|y_i - G'_{\hat{w}} p_i\| \frac{\|\ell\|}{\|p_i\|}$$

and from (9.27)

$$\le \sigma_i \lambda_i \|\ell\|, \lambda_i = \max\{\|w_{i+1} - \hat{w}\| , \|w_i - \hat{w}\|\}. \tag{9.33}$$

Denote

$$\phi_i^2 = \sigma_i(\sigma_i - 2) \frac{\langle p_i, \eta_i\rangle^2}{\langle p_i, p_i\rangle} = \sigma_i(\sigma_i - 2)(\ell((B_i - G'_{\hat{w}})(\frac{p_i}{\|p_i\|})))^2.$$

Hence with (9.32) and (9.33) from (9.31)

$$\|\eta_{i+1}\| \le (\|\eta_i\|^2 - \phi_i^2)^{1/2} + \sigma_i \lambda_i \|\ell\| .$$

Squaring on both sides gives

$$\phi_i^2 \le \|\eta_i\|^2 - \|\eta_{i+1}\|^2 + 2\|\eta_{i+1}\|\sigma_i\lambda_i\|\ell\| + (\sigma_i\lambda_i\|\ell\|)^2,$$

summing from 1 to infinity for some constant $\alpha$

$$\sum_{i=1}^{\infty} \phi_i^2 \le \|\eta_1\| + \sum_{i=1}^{\infty} (2\|\eta_{i+1}\| \sigma_i + \sigma_i^2\lambda_i)\lambda_i. \tag{9.34}$$

(9.29) yields $\sum_{i=1}^{\infty} \lambda_i < \infty$ and the boundedness of $\lambda_i$. Since

$$\|\eta_{i+1}\| \le \|B_i - G'_{\hat{w}}\| ,$$

we obtain from (9.34)

$$\sum_{i=1}^{\infty} \phi_i^2 < \infty$$

which implies

$$0 = \lim_{i\to\infty} \phi_i = \lim_{i\to\infty} \ell((B_i - G'_{\hat{w}})(\frac{p_i}{\|p_i\|})) .$$

REMARK: If E is finite-dimensional, then (9.21) holds and Theorem 9.10 is applicable.

## 9.7 COMMENTS

Section 9.1: The algorithm extending Newton's method to these problems is often called the Osborne-Watson algorithm [3], [4]. Using a Broyden update instead of the inverse Hessian for this method in finite dimensional spaces has been investigated in [5] and [6].

Section 9.2: Strong uniqueness is a very useful tool in the theory of Chebyshev approximation [2].

Section 9.3: Cromme [7] gives a proof of quadratic convergence for the original Osborne-Watson algorithm. Other details of constrained approximation problems are discussed in [8].

Section 9.4: This section particularly shows the parallelism to Chapter 5 when strong uniqueness is present.

Section 9.5: Since linear convergence is usually proved before considering superlinear rate, Theorem 9.10 gives a sufficient condition on the updates.

Section 9.6: Madsen [4] proves superlinear convergence for finite dimensional problems and a Broyden update. He also obtains global convergence using an appropriate step length rule. For a unifying treatment of global convergence and a recent list of references on these methods see Reemtsen [9]. Algorithms of this type have been successfully applied to non-trivial approximation problems in mathematics, such as the calculation of free boundaries in parabolic equations [10], [11].

# 10 Nonlinear parabolic control problems

## 10.1 PARABOLIC DIFFERENTIAL EQUATION WITH A NONLINEAR BOUNDARY CONDITION

We consider a one dimensional diffusion process where the control acts on the boundary and, in addition, the boundary condition is nonlinear. This situation occurs, for example, if the heat transfer at the boundary is not governed by Newton's law, a linear relation, but rather by convection or radiation, i.e. the Stefan and Boltzmann condition. For a discussion of these various laws see the classical book of Carslaw and Jaeger [1]. Nonlinear boundary conditions do not only occur in heat conduction processes but also in chemical diffusion processes, e.g. the Michaelis-Menten law.

Therefore, we deal with the following equations, where $y(t,x)$ denotes the temperature or concentration of the diffusion process at time $t \in [0,T]$ and at point $x \in [0,1]$.

The diffusion process is described by

$$y_t(t,x) = y_{xx}(t,x) \qquad t \in (0,T), \quad x \in (0,1) , \tag{10.1}$$

with initial condition

$$y(0,x) = w(x) \qquad x \in (0,1) , \tag{10.2}$$

homogeneous linear left boundary condition

$$y_x(t,0) = 0 \qquad t \in [0,T] , \tag{10.3}$$

and nonlinear right boundary condition with control function u

$$y_x(t,1) = g(y(t,1)) + u(t) \qquad t \in [0,T] . \tag{10.4}$$

The nonlinear function $g : \mathbb{R} \to \mathbb{R}$ can take various forms. In the cases mentioned above, we have

$$y_x(t,1) = - y(t,1) + u(t) \qquad \text{Newton's law,}$$

$$y_x(t,1) = - y^{5/4}(t,1) + u(t) \qquad \text{convection,}$$

$$y_x(t,1) = -y^4(t,1) + u(t) \qquad \text{Stefan-Boltzmann's law,}$$

$$y_x(t,1) = \frac{y(t,1)}{1+y(t,1)} + u(t) \qquad \text{Michaelis-Menten's law.}$$

We define a solution of the initial boundary value problem (10.1)-(10.4) via the corresponding integral equation.

Let $G(t - \tau, x, \xi)$ be the Green's function which corresponds to the one dimensional heat equation with Neumann boundary conditions. Then a solution satisfies the following integral equation, if w and u are sufficiently smooth:

$$y(t,x) = \int_0^t G(t - s, x, 1)(g(y(s,1)) + u(s))ds + \int_0^1 G(t, x, \xi)\, w(\xi)d\xi \,. \tag{10.5}$$

(10.5) is a nonlinear integral equation for each given function u and w. We could take both or each of them as control functions. In either case the solution y does not depend linearly on u or w, if g is nonlinear.

## 10.2 BOUNDARY CONTROL

For simplicity let us assume $w(\cdot) = 0$.

Thus (10.5) is equivalent to an integral equation for $x = 1$ with $z(t) = y(t,1)$,

$$z(t) = \int_0^t G(t - s, 1, 1)(g(z(s)) + u(s))\, ds \,. \tag{10.6}$$

Since z represents the temperature or concentration, we are interested in nonnegative solutions. So we assume the controls u to be nonnegative and suppose $g(0) \geq 0$, which prevents the solutions from becoming negative. This can be seen from a physical argument. If $y(t,1)$ tends to zero, than as soon as $y(t,1) = 0$ we have

$$y_x(t,1) \geq 0 + u(t) \geq 0 \,,$$

i.e. heat enters the system or at least there is no heat exchange. Hence the temperature cannot drop anymore. Since we need uniqueness of solutions, otherwise we might have a multi-valued cost function, we impose either Lipschitz continuity or the requirement that the nonlinearity g is nonincreasing. Although the solution exists on some interval, there can be a finite escape time smaller than T. Therefore, we assume $u \in L_\infty [0,T]$ and that g lies below an affine-linear functional.

For a proof of the following theorem see Sachs [2]

THEOREM 10.1: Let $g : [0,\infty] \to \mathbb{R}$ be continuous and fulfill for some $\alpha, \beta > 0$

$$g(\eta) \le \alpha + \beta\eta \qquad \text{for all } \eta \ge 0 .$$

Furthermore let g be either nonincreasing on $[0,\infty]$ or locally Lipschitz continuous on $[0,\infty]$. If $g(0) \ge 0$, then there exists for each $u \in L_\infty [0,T]$ with $u(t) \ge 0$ a.e. on $[0,T]$ a unique continuous solution of (10.1) - (10.4), i.e. (10.5) holds with

$$y(t,x) \ge 0 \qquad \text{on} \qquad [0,T] \times [0,1] .$$

The cost function should depend on the distribution of y on the interval $[0,1]$ at time $T > 0$. Thus we have to prove Fréchet differentiability of the following operator

$$S : L_\infty [0,T] \to C [0,1] ,$$

where for $u \in L_\infty [0,T]$, $u \ge 0$

$$Su = y(u ; T, \cdot) , \tag{10.7}$$

$y(u; t,x)$ representing the solution of (10.1)-(10.4) for u. Since the image Su is given by a nonlinear equation, the natural way to prove differentiability is to use the implicit function theorem.

The result proven in [3] for the Stefan-Boltzmann law generalizes easily to differentiable g.

THEOREM 10.2: Let the assumptions of Theorem 10.1 hold and g be differentiable on $(-\varepsilon,\infty)$, for some $\varepsilon > 0$. For each $\hat{u} \in L_\infty [0,T]$, $\hat{u} \ge 0$, the operator S defined by (10.7) is Fréchet differentiable at $\hat{u}$ and $S'_{\hat{u}}(u)$ is given by

$$S'_{\hat{u}}(u) = \int_0^T G(T - s, x, 1)(g'(\hat{z}(s))h(s) + u(s))ds ,$$

where $\hat{z}$ is the solution of (10.6) with $\hat{u}$ and h the unique solution of the linearized equation of (10.6)

$$h(t) = \int_0^t G(t - s, 1, 1)(g'(\hat{z}(s))h(s) + u(s))ds .$$

Keeping in mind the definition of a solution via integral equations, the derivative $S'_{\hat{u}}(u)$ can also be interpreted as $d(T,\cdot)$ where d solves

$$
\begin{aligned}
d_t(t,x) &= d_{xx}(t,x) && t \in (0,T),\ x \in (0,1),\\
d(0,x) &= 0 && x \in (0,1),\\
d_x(t,0) &= 0 && t \in [0,T] ,\\
d_x(t,1) &= g'(y(\hat{u};\ t,\ 1))d(t,1) + u(t) && t \in [0,T] .
\end{aligned}
$$

If we suppose that g is locally Lipschitz continuous on $[0,\infty]$, then we can also show that $S'_{\hat{u}}$ is Lipschitz continuous in a neighborhood of $\hat{u}$.

## 10.3 INITIAL CONTROL

In this section we set $u \equiv 0$. Another way to prove the Fréchet differentiability in (10.1) - (10.4) is to start with the optimal control $\hat{w}$ and apply the implicit function theorem. Then there exists a solution operator in a neighborhood of $\hat{w}$ which is differentiable.

The integral equation (10.5) reduces with $z(t) = y(t,1)$ to

$$z(t) = \int_0^t G(t - s, 1, 1)g(z(s))ds + \int_0^1 G(t, 1, \xi)w(\xi)d\xi . \qquad (10.8)$$

THEOREM 10.3: Let $\hat{w} \in L_2\,[0,1]$ be an optimal control and $\hat{z} \in C[0,1]$ the corresponding solution of (10.8). If $g : \mathbb{R} \to \mathbb{R}$ is differentiable with locally Lipschitz continuous derivative, then there exists a continuous map R from a neigborhood $N(\hat{w})$ of $\hat{w}$ into $C\,[0,1]$, which is differentiable at $\hat{w}$ and has the following properties:

(i) $R\hat{w} = \hat{z}$,

(ii) Rw solves (10.8) for each $w \in N(\hat{w})$,

(iii) The derivative $R'_{\hat{w}}$ given by

$R'_{\hat{w}}\, w = h$ solves

$$h(t) = \int_0^t G(t - s, 1, 1)\, g'(\hat{z}(s))\, h(s)ds + \int_0^1 G(t, 1, \xi)\, w(\xi)d\xi \;. \tag{10.9}$$

Sketch of the proof: We define a map

$$A : C\,[0,T] \times L_2\,[0,1] \to C\,[0,T]$$

by

$$A(a,b) = -\,a(t) + \int_0^t G(t - s, 1, 1)g(a(s))ds + \int_0^1 G(t, 1, \xi)\, b(\xi)d\xi \;.$$

This map is Fréchet differentiable with respect to $\hat{a}$ and $\hat{b}$ with continuous derivatives $A'_{\hat{a}}$, $A'_{\hat{b}}$. $A'_{\hat{a}}$ is bijective if and only if

$$a(t) = \int_0^t G(t - s, 1, 1)\, g'(\hat{a}(s))\, a(s)ds$$

is only trivially solvable, which is true for linear integral equations with weakly singular kernels (Michlin [4]). From the implicit function theorem we obtain (i), (ii) and that the derivative $R'_{\hat{w}}$ of $R$ at $\hat{w}$ is given by

$$R'_{\hat{w}} = -\,[A'_{\hat{a}}]^{-1}\, A'_{\hat{b}} \;,$$

i.e. $h = R'_{\hat{w}}\, w$ satisfies (10.9).

Similarly, we can show the differentiability in the neighborhood $N(\hat{w})$ and hence the Lipschitz continuity of $R'_w$ in $N(\hat{w})$.

Since we want $y(T,\cdot)$, as in the previous section, to enter the cost function, we apply another map

$$z \to \int_0^T G(T - s, x, 1)g'(z(s))ds + \int_0^1 G(T, x, \xi)w(\xi)d\xi \;.$$

which can easily be shown to be differentiable. Hence the derivative of the operator

$$\overline{S} : w \to y(w;\, T,\cdot)$$

is given by the solution of

$$d(t,x) = \int_0^t G(t-s, x, 1)\, g'(y(\hat{u}; s, 1))\, d(s,1)\, ds + \int_0^1 G(t, x, \xi)\, w(\xi) d\xi ,$$

or written as the initial boundary value problem

$$\begin{aligned} d_t(t,x) &= d_{xx}(t,x) && t \in (0,T), \quad x \in (0,1), \\ d(0,x) &= w(x) && x \in [0,1], \\ d_x(t,0) &= 0 && t \in [0,T], \\ d_x(t,1) &= g'(y(\hat{u}; t, 1)d(t,1) && t \in [0,T]. \end{aligned}$$

## 10.4 OPTIMAL CONTROL PROBLEMS

After our discussion of a few analytical properties of the systems, we now formulate the optimization problem. Here we want to consider problems of reaching a certain given distribution of temperature or concentration $a(x)$, $x \in [0,1]$.

As two examples, we minimize

$$\int_0^1 (y(u; T, x) - a(x))^2 dx$$

or

$$\max_{0 \le x \le 1} |y(u; T, x) - a(x)| .$$

We have seen in Section 10.2 that u was restricted to be non-negative. Since also the controls, i.e. heat input, must be bounded from a physical point of view, we introduce the following set of admissible controls,

$$U = \{u \in L_\infty [0,T] : 0 \le u(t) \le 1, \quad t \in [0,T] \quad \text{a.e.}\} .$$

Hence we have problem (P 1):

Find $\hat{u} \in U$ such that for all $u \in U$ (P 1)

$$\|y(\hat{u}; T,\cdot) - a\|_\infty \le \|y(u; T,\cdot) - a\|_\infty .$$

Another possibility to avoid controls whose $L_2$-norms become unbounded is to define problem (P 2):

Find $\hat{u} \in L_2\ [0,T]$ such that for all $u \in L_2\ [0,T]$ (P 2)

$$\|y(\hat{u};\ T,\cdot) - a\|_2 + \|\hat{u}\|_2 \leq \|y(u;\ T,\cdot) - a\|_2 + \|u\|_2 .$$

We could also replace u by w and obtain an initial control problem. These are only two examples from a large variety of objectives. In the sequel we will discuss some of the remaining requirements for the convergence statements. We begin by checking second derivative properties for the maximum-norm case (P 1).

Two conditions have to be assured, injectivity of the derivative at the optimal point and some type of strong uniqueness property. In order to put our examples into the framework of Chapter 9, we choose a discretization of U:

$$U_n = \{ \sum_{i=1}^{n} v_i \chi_i \in L_\infty\ [0,T] : v_i \in [0,1],\ \ i = 1,\ldots,n\} ,$$

where $\chi_i$, $i = 1,\ldots,n$, are the characteristic functions on $[t_{i-1}, t_i]$, $0 = t_0 < t_1 < \ldots < t_n = T$.

Then problem (P 1) can be replaced by (P 3).

Find $\hat{v} \in U_n$ such that for all $v \in U_n$ (P 3)

$$\|y(\hat{v};\ T,\cdot) - a\|_\infty \leq \|y(v;\ T,\cdot) - a\|_\infty .$$

Here, $y(v;\ t,\ x)$ denotes the solution of the differential equation with control function

$$\sum_{i=1}^{n} v_i \chi_i(t) .$$

As seen in Sections 9.3 - 9.5, the derivative of the nonlinear approximating family of functions with respect to the parameter was required to be injective.

This means for the boundary control problem, if d is a solution of

$$\begin{aligned}
&d_t(t,x) = d_{xx}(t,x) && t \in (0,T),\ x \in (0,1),\\
&d_x(t,0) = 0 && t \in (0,T),\\
&d_x(t,1) = g'(y(\hat{v};\ t,\ 1))d(t,1) + u(t) && t \in (0,T),\\
&d\ (T,x) = 0 = d(0,x) && x \in [0,1],
\end{aligned}$$

for some $u \in U_n$, then $u \equiv 0$ or equivalently $d \equiv 0$. Let us discuss this problem for the linear case. If $g' \equiv -c$, $c > 0$, then Green's function can be expressed as a series

$$G(t, x, \xi) = \sum_{k=1}^{\infty} \exp(-\lambda_k t)\, e_k(x)\, e_k(\xi) ,$$

with $\lambda_k$ and $e_k$ as the eigenvalues and eigenfunctions, respectively, of the corresponding Sturm-Liouville problem. Hence

$$0 \equiv d(T, x) = \sum_{k=1}^{\infty} e_k(x) \int_0^T \exp(-\lambda_k t)\, u(t)\, dt\, e_k(1) ,$$

and since the eigenfunctions are dense in $C[0,1]$,

$$\int_0^T \exp(-\lambda_k t)\, u(t) dt = 0 \quad \text{for all } k \in \mathbb{N} . \tag{10.10}$$

The eigenvalues are such that

$$\sum_{k=1}^{\infty} \lambda_k^{-1} < \infty ,$$

therefore the Theorem of Müntz implies that

$$\{\exp(-\lambda_k t)\}_{\mathbb{N}} \quad \text{is not dense in } C[0,1] .$$

Proofs of these results can be found in Michlin [5] . Thus, there is a function $u \in L_2[0,T]$ which satisfies (10.10) and by these arguments we cannot deduce that u must be zero. Although the functions v in this problem are from a smaller (finite dimensional) space $U_n$, there is a time-dependent coefficient in the boundary condition that destroys the applicability of this reasoning. Hence, the proof of injectivity is not a completely trivial problem.

If we consider the initial value control problem, then the injectivity condition implies:

If d is a solution of

$$\begin{aligned}
&d_t(t,x) = d_{xx}(t,x) && t \in (0,T), \quad x \in (0,1), \\
&d(0,x) = w(x) && x \in [0,1], \\
&d_x(t,0) = 0 && t \in (0,T), \\
&d_x(t,1) = g'(y(\hat{w}; t, 1))d(t,1) && t \in (0,T), \\
&d(T,x) = 0 && x \in [0,1],
\end{aligned}$$

for some $w \in L_2 [0,1]$, then necessarily $w \equiv 0$.

This result is well known in the theory of parabolic differential equations under the name of the backward uniqueness property. If g is linear, then this property holds. For time-dependent boundary conditions the theorems on backward uniqueness are closely connected with the concept of a solution of a differential equation.

## 10.5 STRONG UNIQUENESS

The second requirement we have to deal with is the strong uniqueness property. In Lemma 9.6 we have seen already how the strong uniqueness can be obtained if there are no restrictions on the parameters.

Before discussing the details of the application to our models, we review some extensions of this lemma to the constrained case. We define a generalization of a Haar's space of functions.

DEFINITION 10.4: Let $f_1,\dots,f_1 \in C(I)$, I compact interval, $m \in \mathbb{N}$, $m < n$. Then the cone

$$C = \{ \sum_{i=1}^{n} \alpha_i f_i : \alpha \in \mathbb{R}^n, \quad \alpha_i \geq 0 \qquad i = m+1,\dots,n\}$$

is called a Haar's cone if and only if for each subset J of $\mathbb{N}$ with

$$\{1,2,\dots,m\} \subseteq J \subseteq \{1,2,\dots,n\}$$

the functions $\{f_j : j \in J\}$ form a Haar's space.

Then we can prove the following

THEOREM 10.5: Let $F : V \to C(I)$ be Fréchet differentiable at $\hat{v} \in V$,

$$V = \{v \in \mathbb{R}^n : 0 \leq v_i \leq 1, \quad i = 1,\dots,n\} ,$$

$F'_{\hat{v}}$ injective and $\hat{v}$ a best approximation, i.e.

$$\|F(\hat{v})\| \leq \|F(v)\| \qquad \text{for all } v \in V ,$$

$\{e_i\}_{i=1}^{n}$ the set of canonical unit vectors, and

$$J_- = \{i \in \mathbb{N} : \hat{v}_i = 0\}, \quad J_+ = \{i \in \mathbb{N} : \hat{v}_i = 1\} .$$

If the cone

$$C = \{ \sum_{i=1}^{n} \alpha_i F'_{\hat{v}}(e_i) : \alpha \in \mathbb{R}^n , \quad \alpha_i \geq 0 \quad \text{for} \quad i \in J_- , \quad \alpha_i \leq 0 \quad \text{for} \quad i \in J_+ \}$$

is a Haar's cone, then $F(\hat{v})$ is a locally strongly unique best approximation.

Sketch of proof: If $\hat{v}$ is a best approximation, then

$$\|F(\hat{v})\| \leq \|F(\hat{v}) - F'_{\hat{v}}(h)\| \tag{10.11}$$

for all $h \in T(V,\hat{v}) = \{h \in \mathbb{R}^n : h_i \geq 0$ for $i \in J_-$, $h_i \leq 0$ for $i \in J_+\}$, see e.g. Collatz-Krabs [6].

Braess [7] has shown that, if (10.11) holds and $F'_{\hat{v}}(T(V,\hat{v})) = C$ is a Haar's cone, then $\Theta$ is strongly unique best approximation of $F(\hat{v})$ in C. This however implies with Wulbert [8] that $F(\hat{v})$ is a locally strongly unique best approximation.

Since the sets $J_-$ and $J_+$ are not a priori known, we have to show the following:

Denote $f_i$ by

$$f_i = F'_{\hat{v}}(e_i) .$$

The requirement concerning the Haar's cone in Theorem 10.5 is satisfied, if for any subset J of $\{1,\dots,n\}$ and any $\alpha \in \mathbb{R}^{|J|}$, the function

$$\sum_{i \in J} \alpha_i f_i$$

vanishes identically or has at most $|J| - 1$ zeroes.

In order to prove this for our parabolic problems, we cite a theorem of Karafiat [9].

THEOREM 10.6: Let c be a solution of

$$\begin{aligned} c_t(t,x) &= c_{xx}(t,x) && t \in (0,T], \quad x \in (-1,1), \\ c(0,x) &= g(x) && x \in [-1,1], \\ -c_x(t,-1) &= \beta(t)\, c(t,-1) + u(t) && t \in (0,T], \\ c_x(t,1) &= \gamma(t)\, c(t,1) + v(t) && t \in (0,T], \end{aligned}$$

$\beta, \gamma \in C(0,T]$, non-negative, g bounded, piecewise continuous, continuous at $\pm 1$ and continuous from the left on $[-1,+1]$, $u, v \in L_\infty [0,T]$.

Suppose u and v have p and r switches on (0,T), respectively, and g changes sign n times on [-1,1]. If the function c(T,x) of the variable x changes sign on (-1,1) $m_1$ times and has $m_2$ zeroes at which it does not change sign; then

(i) $m_1 + 2m_2 \leq n + p + r + 2$ ,

(ii) if a and b are the first switching points of u and v respectively, $g \equiv 0$, and

$$\int_0^a u(t)dt \int_0^b v(t)dt > 0 ,$$

then $m_1 + 2m_2 \leq p + r$,

(iii) if $u \equiv v \equiv 0$ , then

$$m_1 + 2m_2 \leq p + r.$$

We have to define the various concepts of sign changes and zeroes:

<u>DEFINITION 10.7</u>: A function f defined on an interval $(\alpha,\beta)$ changes sign n times if and only if

$$n = \sup \{m : \exists\ \alpha < x_1 < x_2 < \ldots < x_{m+1} < \beta :$$
$$f(x_i)\ f(x_{i+1}) < 0 \quad \text{for} \quad i = 1,\ldots,m\} .$$

The function $f \in C(\alpha,\beta)$ has n zeroes at which it does not change sign if and only if

$$n = \sup \{m : \exists\ \alpha < x_1 < x_2 < \ldots x_{3m} < \beta : \text{ for } i = 3k - 1,$$
$$k = 1,\ldots,m,\ f(x_i) = 0,\ f(x_{i-1})\ f(x_{i+1}) > 0 \quad \text{and}$$
$$f(x')\ f(x") \geq 0 \quad \text{for all} \quad x', x" \in (x_{i-1}, x_{i+1})\} .$$

The integrable function f on $(\alpha,\beta)$ has n switches on the interval $(\alpha,\beta)$ if and only if n is the smallest number such that there are points $\alpha = x_0 < x_1 < \ldots < x_{n+1} = \beta$ for which the function f has a constant sign almost everywhere on each interval

$(x_i, x_{i+1})$, $i = 0,\dots,n$, i.e. for almost all $x', x'' \in (x_i, x_{i+1})$ we have $f(x')\, f(x'') \geq 0$.

In order to apply Theorem 10.5 we have to assure that $c(T,\cdot)$ has only finitely many zeroes, which is not evident if we have time-dependent coefficients in the boundary condition. Similarly, statements on the zeroes of $c(T,\cdot)$ do not include the endpoints of the interval. Therefore, we perturb the approximation problem (P1):

Instead of

$$y(T,x), \quad x \in [0,1] ,$$

we observe

$$y(T + \delta, x), \quad x \in [0, 1 - \delta] ,$$

and minimize

$$\max_{0 \leq x \leq 1 - \delta} |y\,(u;\, T + \delta,\, x)| \quad \text{over} \quad u \in U .$$

The extended boundary condition on $[T, T + \delta]$ is

$$\left.\begin{aligned} y_x(t,0) &= 0 \\ y_x(t,1) &= g'(y(\hat{u};T,1))\,y(t,1) , \end{aligned}\right\} \quad t \in [T, T + \delta], \tag{10.12}$$

where $\hat{u}$ is the optimal control.

This perturbation is only for theoretical purposes, because (10.12) gives a continuous extension of the linearized system.

<u>THEOREM 10.8:</u> If $g'(y(\hat{u};t,1)) \leq 0$ for $t \in [0,T]$, then the perturbed boundary control problem satisfies the strong uniqueness condition for any $\delta > 0$.

<u>Proof:</u> By Theorem 10.2 the Fréchet derivative of the unperturbed problem is given by $d(T,x)$ solving

$$d(t,x) = \int_0^t G(t - s,\, x,\, 1)\,(g'(y(\hat{u};s,1))d(s,1) + u(s)ds .$$

We extend d symmetrically to the left side of 0, and since $u \in U_n$ is piecewise continuous we can write d as a solution of the following differential equation:

$$\begin{aligned}
d_t(t,x) &= d_{xx}(t,x) && t \in (0, T+\delta),\ x \in (-1,1),\\
d(0,x) &= 0 && x \in (-1,1),\\
-d_x(t,-1) &= \gamma(t)d(t,-1) + \overline{u}(t) && \\
d_x(t,1) &= \gamma(t)d(t,1) + \overline{u}(t) && t \in (0, T+\delta),
\end{aligned}$$

where

$$\gamma(t) = \begin{cases} g'(y(\hat{u};t,1)) & t \in [0,T), \\ g'(y(\hat{u};T,1)) & t \in [T, T+\delta], \end{cases}$$

$$u(t) = \begin{cases} \sum_{i \in |J|} v_i \chi_i(t) & t \in [0, T), \\ 0 & t \in [T, T+\delta], \end{cases}$$

$J$ some subset of $\{1,\ldots,n\}$.

Since $d$ for $t \in [T, T+\delta]$ is solution of a parabolic differential equation with constant coefficients and homogeneous boundary condition, $d(T,\cdot)$ is analytic and vanishes or has only finitely many zeroes. Since $\gamma$ is nonpositive and continuous, we can apply Theorem 10.6(ii) yielding

$$m_1 + 2m_2 \le p + r .$$

Since $u$ is a linear combination of $|J|$ characteristic functions, the numbers of switches $p$ and $r$ are estimated by

$$p = r \le |J| - 1 .$$

Hence

$$m_1 + 2m_2 \le 2(|J| - 1) ,$$

i.e. the number of zeroes in $(-1,1)$ is less them $2(|J| - 1) - m_2$. Since in $x = 0$ only a zero without sign change can occur, we derive that $d(T,\cdot)$ has at most $|J| - 1$ zeroes on $[0,1)$. This statement holds for any $\delta > 0$.

<u>THEOREM 10.9:</u> If $g'(y(\hat{u};t,1) \le 0$ for $t \in [0,T]$, then for any $\delta > 0$ the perturbed initial value control problem satisfies the

strong uniqueness property.

The proof follows that of the previous theorem.

The condition $g'(y(\hat{u};t,1)) \leq 0$ for $t \in [0,T]$ can be satisfied easily by assuming in Theorem 10.1 that g is decreasing, which is true for the convection and radiation boundary condition.

## 10.6 COMMENTS

Section 10.1: Some of the cited problems are suggested in [10].

Section 10.2: Parabolic control problems with nonlinear boundary conditions are treated in [3] amd [2] where more references can be found.

Section 10.3: A maximum principle for various types of controls and cost functionals is derived in [11].

Section 10.4: Backward uniqueness properties in optimal control are also discussed in [12]. Approximation theory is treated by Cheney [9]. For a treatment of observability and controllability, which is closely connected to the contents of this section, see [14] for example.

Section 10.5: For the proof of strong uniqueness of a parabolic free boundary problem, see [15]. A different approach to the solution of parabolic control problem with nonlinear boundary is presented in [16].

# 11 Linear parabolic control problems

## 11.1 GENERAL ASPECTS OF THE BANG-BANG PRINCIPLE

We shall consider linear parabolic control problems with controls lying in a feasible set. Intuitively, it is clear that the optimal controls will be found on the boundary of this set. This gives a particular representation for optimal controls which will be used partially in designing numerical procedures for solving these problems. Therefore, we give a short introduction into the bang-bang principle for linear problems.

Let us consider the following problem: $T > 0$,

$$U = \{u \in L_\infty[0,T] : |u(t)| \le 1 \text{ a.e. in } [0,T]\}$$

$$S : L_p[0,T] \to G ,$$

for some $p \in [1,\infty]$, $G$ Banach space, $z \in G$ fixed,

$$\psi : G \to \mathbb{R} \qquad \text{convex, continuous .}$$

The general problem consists of finding $\hat{u} \in U$ with

$$\psi(S\hat{u} - z) \le \psi(Su - z) \qquad \text{for all } u \in U . \tag{11.1}$$

Furthermore we assume, that the adjoint operator $S^*$ fulfills

$$S^*(G) \subseteq L_q[0,T] , \tag{11.2}$$

$p^{-1} + q^{-1} = 1$ for $p < \infty$ and $q = 1$ for $p = \infty$ .

Since $\psi$ is convex, we have the following characterization of optimal points:

THEOREM 11.1: The control $\hat{u} \in U$ is optimal if and only if there exists

$$\hat{\ell} \in \partial\psi(S\hat{u} - z) \tag{11.3}$$

such that

$$|\hat{u}(t)| = 1 \tag{11.4}$$

for almost all $t \in [0,T]$ with $S^*\hat{\ell}\,(t) \neq 0$.

For a proof we refer to [1].

(11.3) is by definition of the subgradient equivalent with

$$\hat{\ell} \in G^*, \quad \hat{\ell}(S\hat{u} - z) \leq \hat{\ell}(g) \qquad \text{for all } g \in G . \tag{11.5}$$

Hence the problem of (11.4) is to determine the set of zeros of $S^*\hat{\ell}$. There are various forms of bang-bang principles:

<u>DEFINITION 11.2:</u> If for each optimal control $\hat{u} \in U$

a) there is a measurable set $I \subseteq [0,T]$ with Lebesgue measure $\mu(I) > 0$ such that

$$|\hat{u}(t)| = 1 \quad \text{a.e. on } I , \tag{11.6}$$

then a <u>weak</u> <u>bang-bang</u> <u>principle</u> holds.

b)

$$|\hat{u}(t)| = 1 \quad \text{a.e. on } [0,T] , \tag{11.7}$$

then a <u>strong</u> <u>bang-bang</u> <u>principle</u> holds.

c) there are

$$\{t_i\}_{\mathbb{N}} , \quad t_i \in [0,T]$$

such that for $\varepsilon = 1$ or $\varepsilon = -1$ and all $i \in \mathbb{N}$

$$\hat{u}(t) = \varepsilon(-1)^i \quad \text{a.e. on } [t_i, t_{i+1}) , \tag{11.8}$$

then a <u>countable</u> <u>bang-bang</u> <u>principle</u> holds.

d) there are finitely many

$$\{t_i\}_{i=1}^m , \quad t_i \in [0,T]$$

such that for $\varepsilon = 1$ or $\varepsilon = -1$ and all $1 \leq i \leq m$

$$\hat{u}(t) = \varepsilon(-1)^i \quad \text{a.e. on } [t_i, t_{i+1}) ,$$

then a <u>finite</u> <u>bang-bang</u> <u>principle</u> holds.

If $\hat{u} \in U$, an optimal point, is also optimal for the unconstrained problem, then we cannot expect $\hat{u}$ to satisfy a bang-bang principle, since the bound on $\hat{u}$ is redundant. Therefore we define,

DEFINITION 11.3: Let A be a nonvoid set of G. Then $S\hat{u} \in A$ is called A-minimal if and only if

$$\psi(S\hat{u} - z) \le \psi(a - z) \qquad \text{for all } a \in A. \tag{11.9}$$

Each image of a solution of (11.1) is S(U)-minimal, but not necessarily $S(L_p)$-minimal or even G-minimal.

THEOREM 11.4: If there is some solution $\hat{u} \in U$ of (11.1) such that $S\hat{u}$ is not $S(L_p)$-minimal, then the weak bang-bang principle holds.

Proof: Let $u^* \in U$ satisfy

$$\psi(Su^* - z) \le \psi(Su - z) \qquad \text{for all } u \in U \ .$$

Since $S\hat{u}$ is not $S(L_p)$-minimal, $Su^*$ is also not $S(L_p)$-minimal. By a well known characterization of optimal points, see e.g. [2],

$$\Theta \notin S^*(\partial\psi(Su^* - z)) \ .$$

Hence for $\ell^*$ determined by (11.3)

$$S^*\ell^* \neq \Theta \ ,$$

i.e. $|\hat{u}(t)| > 1$ on a set of measure greater zero.

In order to prove strong bang-bang principles, we require more information on the adjoint operator:

DEFINITION 11.5: A linear operator $S : U \to G$ is called normal if and only if for each

$$u \in U \ , \quad \ell \in \partial\psi(Su - z) \ , \ S^*\ell \neq \Theta \ , \tag{11.10}$$

we have that the set

$$N = \{t \in [0,T] : S^*\ell(t) = 0\} \text{ is of measure zero} \ . \tag{11.11}$$

S is called countably (finitely) normal, if N has only a countable (finite) number of elements.

THEOREM 11.6: Let S be (countably, finitely) normal. If there is a solution $\hat{u} \in U$ of (11.1) such that $S\hat{u}$ is not $S(L_p)$-minimal, then the strong (countable, finite) bang-bang principle holds.

The proof is evident from Theorem 11.4 and Definition 11.5.

In a numerical procedure, the operator S and z are often approximated by $S_n$ and $z_n$. Therefore, let for $n \in \mathbb{N}$

$$S_n : L_p[0,T] \to G , \quad z_n \in G$$

be given and find $\hat{u}_n \in U$ such that

$$\psi(S_n\hat{u}_n - z_n) \leq \psi(S_n u - z_n) \quad \text{for all } u \in U . \tag{11.12}$$

The question arises whether bang-bang principles also hold for these problems (11.12).

<u>THEOREM 11.7:</u> Let

$$\lim_{n\to\infty} z_n = z \quad \text{and} \quad \lim_{i\to\infty} S_n = S \tag{11.13}$$

in the operator norm, $\{S_n\}_{\mathbb{N}}$ be compact and normal operators. If there is a solution $\hat{u} \in U$ of (11.1) such that $S\hat{u}$ is not $S(L_p)$-minimal, then there is $n_o \in \mathbb{N}$ and the strong bang-bang principle holds for all problems (11.12) with $n \geq n_o$.

<u>Proof:</u> According to Theorem 11.6 we only have to prove that there exists $n_o \in \mathbb{N}$ such that for each $n \geq n_o$ there are solutions $u_n$ of (11.12) which have the property that $S_n u_n$ is not $S_n(L_p)$-minimal. If this is not true, then there is a subsequence $\{n_k\} \subseteq \mathbb{N}$ and solutions $\hat{u}_{n_k}$ of (11.12) which satisfy even

$$\psi(S_{n_k}\hat{u}_{n_k} - z_{n_k}) \leq \psi(S_{n_k} u - z_{n_k}) \tag{11.14}$$

for all $u \in L_p[0,T]$. Suppose for a moment, that for some $\hat{u} \in U$

$$\lim_{i\to\infty} \psi(S_{n_k}\hat{u}_{n_k} - z_{n_k}) = \psi(S\hat{u} - z) \tag{11.15}$$

holds. Then from (11.14) and the convergence properties of $S_n$ and $z_n$

$$\psi(S\hat{u} - z) \leq \psi(Su - z) \quad \text{for all } u \in L_p[0,T] ,$$

i.e. $S\hat{u}$ is $S(L_p)$-minimal, a contradiction.

$\{\hat{u}_{n_k}\}_{\mathbb{N}}$ is a subsequence converging weakly (*) to some $\overline{u} \in U$.

$$\psi(S_{n_k}\hat{u}_{n_k} - z_{n_k}) \leq \psi(S_{n_k}\overline{u} - z_{n_k})$$

implies

$$\limsup_{k \to \infty} \psi(S_{n_k} \hat{u}_{n_k} - z_{n_k}) \le \psi(S\bar{u} - z) \ .$$

Assume, there exists a subsequence also called $\{n_k\}$ and $\varepsilon > 0$ such that

$$\psi(S_{n_k} \hat{u}_{n_k} - z_{n_k}) \le \psi(S\bar{u} - z) - \varepsilon \ , \quad k \in \mathbb{N} \ . \tag{11.16}$$

Recall, $\{S_n\}_{\mathbb{N}}$ are compact operators and with (11.13) S is also compact. Then we can find a subsequence, also called $\{n_k\}$, such that for some $a \in S(U)$

$$\lim_{k \to \infty} S\hat{u}_{n_k} = a \ .$$

(11.16) implies

$$0 < \varepsilon \le \psi(S\bar{u} - z) - \psi(S_{n_k} \hat{u}_{n_k} - z_{n_k})$$

$$\le \psi(S\hat{u}_{n_k} - z) - \psi(S_{n_k} \hat{u}_{n_k} - z_{n_k}) \ . \tag{11.17}$$

Since the operator convergence yields

$$\lim_{k \to \infty} S_{n_k} \hat{u}_{n_k} = a \ ,$$

taking the limit in (11.17) gives a contradiction. Hence (11.15) is true and the proof is completed.

## 11.2 OPTIMAL CONTROL OF THE HEAT EQUATION

We consider a process which can be described by the heat equation:

Let $\Omega$ be a bounded, open domain with $C^\infty$-boundary $\partial\Omega$, let $\Delta$ denote the Laplacian operator on $\mathbb{R}^n$, and let $\frac{\partial}{\partial \nu}$ denote differentiation with respect to the outward pointing normal $\nu$ of $\partial\Omega$.

If y denotes the temperature of a medium at point $x \in \Omega$ and at time $t \in [0,T]$, then the heat equation with a heat source f holds

$$y_t(t,x) = \Delta y(t,x) + f(t,x), \ t \in (0,T), \quad x \in \Omega \ , \tag{11.18}$$

which describes the variation of the temperature in dependence of t and x.

The initial temperature distribution in $\Omega$ is given by

$$y(0,x) = w(x) \ , \quad x \in \Omega. \tag{11.19}$$

Let the boundary conditions be denoted by

$$a \frac{\partial y}{\partial \nu} (t,\xi) + y(t,\xi) = g(t,\xi), \quad t \in (0,T], \quad \xi \in \partial\Omega \ , \tag{11.20}$$

with some $a \geq 0$.

There are three basic types of control inputs into the system:

<u>distributed</u> control: $f(t,x)$ ,
<u>boundary</u> control: $g(t,\xi)$ ,
<u>initial</u> control: $w(x)$ .

In addition, various observation structures and objective functions can be imposed:

<u>Final state cost:</u> $f(y(T,\cdot))$ with an objective function

$$f : L(\Omega) \to \mathbb{R} \ ,$$

where $L(\Omega)$ is some appropriate function space such that $y(T,\cdot) \in L(\Omega)$.

<u>Distributed state cost:</u> $f_D(y(\cdot,\cdot))$ with an objective function

$$f_D: L((0,T) \times \Omega) \to \mathbb{R} \ ,$$

where $L((0,T) \times \Omega)$ is a function space to be specified such that $y(\cdot,\cdot) \in L((0,T) \times \Omega)$.

Hence it is easy to compose a variety of optimal control problems with various controls and objective functions.

In the sequel, we discuss the dependence of the state y on the controls f, g and w.

Let $\{v_k\}_{\mathbb{N}} \subseteq L_2(\Omega)$ be the complete orthonormal set of eigenfunctions and $\{-\lambda_i\}_{\mathbb{N}}$ the set of corresponding eigenvalues satisfying

$$\Delta v_k(x) = \lambda_k v_k(x) \;, \qquad x \in \Omega \;,$$

$$\frac{\partial}{\partial \nu} \; v_k(\xi) = 0 \;, \qquad \xi \in \partial\Omega \;.$$

For f, g and w "smooth enough", this will be specified later, we can express the solution of (11.18) - (11.20) by Green's function,

$$y(t,x) = \int_\Omega G(t;x,r)w(r)dr + \int_0^t \int_\Omega G(t - s;x,r)f(s,r)dr\, ds$$
$$+ \int_0^t \int_{\partial\Omega} \overline{G}(t - s;x,\xi)g(s,\xi)d\xi\, ds \;. \tag{11.21}$$

Green's function can be expressed by eigenfunction expansion:

$$G(t;x,r) = \sum_{k=1}^{\infty} \exp\,(-\lambda_k t)v_k(x)v_k(r)$$

$$\overline{G}(t;x,\xi) = \sum_{k=1}^{\infty} \exp\,(-\lambda_k t)v_k(x)\overline{v}_k(\xi) \;,$$

$$\overline{v}_k(\xi) = \begin{cases} \frac{1}{a}\, v_k(\xi) & \text{if} \quad a > 0 \;, \\ -\frac{\partial}{\partial\nu} v_k(\xi) & \text{if} \quad a = 0 \;. \end{cases}$$

Let us confine our attention to the situation where $\Omega = (0,1)$. Then $\{\lambda_k\}_{\mathbb{N}}$ is an increasing sequence tending to infinity asymptotically as $k^2$:

$$\lambda_k = \pi^2(k + \alpha)^2 + O(1) \;, \qquad k \to \infty \;. \tag{11.22}$$

This growth of the eigenvalues allows a few conclusions on the spaces which can be selected. We are mainly dealing with initial and boundary controls and observing the final state $y(T,\cdot)$.

<u>LEMMA 11.8:</u> If y is defined by (11.21), we define an operator S by

$$S(w,\, u_0,\, u_1) = y(T,\cdot) \;,$$

where w is the initial state and $u_0$, $u_1$ are the boundary inputs on the left and right side of the interval. Then S is a continuous operator with respect to the following spaces:

$$S : L_2\,[0,1] \times L_2\,[0,T]^2 \to L_2\,[0,1]\ , \tag{11.23}$$

$$S : L_2\,[0,1] \times L_\infty\,[0,T]^2 \to C\ [0,1]\ . \tag{11.24}$$

For a proof we refer to [ 3 ].

Next we make use of the theorems in Section 11.1. Using (11.23) and (11.24) $S^*$ is given by

$$S^*\ell = \Big( \sum_{k=1}^{\infty} \exp(-\lambda_k T)\, v_k(\cdot)\, \ell(v_k)\ , \tag{11.25}$$

$$\sum_{k=1}^{\infty} \exp(-\lambda_k(T-\cdot))\, \overline{v}_k(0)\, \ell(v_k)\ ,$$

$$\sum_{k=1}^{\infty} \exp(-\lambda_k(T-\cdot))\, \overline{v}_k(1)\, \ell(v_k)\Big)\ ,$$

and $S^*$ satisfies

$$S^* : L_2\,[0,1] \to L_2\,[0,1] \times L_2\,[0,T]^2$$

$$S^* : C\,[0,1]^* \to L_2\,[0,1] \times L_1\,[0,T]^2\ .$$

In order to show a weak bang-bang principle we have to impose a condition that assures that a solution $(\hat{w}, \hat{u}_0, \hat{u}_1)$ is not a solution of the unconstrained problem. Select the cost function $\psi$ as the norms

$$\psi = \|\cdot\|_2 \qquad \text{for (11.23)}\ ,$$

and

$$\psi = \|\cdot\|_\infty \qquad \text{for (11.24)}\ .$$

We consider the problem of minimizing

$$\psi(S(w, u_0, u_1))$$

on the whole space. The images

$$S(L_2\,[0,1],\ L_2\,[0,T],\ L_2\,[0,T])$$

$$S(L_2\,[0,1],\ L_\infty\,[0,T],\ L_\infty\,[0,T])$$

are dense in the range space if and only if the adjoint operators are injective.

Concerning the initial control

$$\sum_{k=1}^{\infty} \exp(-\lambda_o T)\, \ell(v_k)\, v_k(x) = 0 \qquad \text{for all } x \in [0,1]$$

implies

$$\ell(v_k) = 0 \qquad \text{for all } k \in \mathbb{N}$$

and with the completeness of $\{v_k\}_{\mathbb{N}}$

$$\ell = \Theta .$$

Similarly for the boundary controls we obtain $\ell = \Theta$, because the functions

$$\{\exp(-\lambda_k(T-\cdot))\}_{\mathbb{N}}$$

in the one dimensional case have the property that

$$\sum_{k=1}^{\infty} \varepsilon_k \exp(-\lambda_k s) \equiv 0 \quad \text{on} \quad [0,T]$$

implies $\varepsilon_k = 0$ for all $k \in \mathbb{N}$.

Since the operators S have dense images, the infimal value of the unconstrained problem is always zero. Hence the assumption in Theorem 11.4 is equivalent with supposing that the minimal value is greater zero.

We confine ourselves to two model problems:

Find

$$\hat{w} \in U_w = \{w \in L_2[0,1] : |w(x)| \le 1 \quad \text{a.e. on } [0,1]\}$$

such that with $u_o = u_1 = \Theta$, $z \in C[0,1]$,

$$\|S\hat{w} - z\| \le \|Sw - z\| \qquad \text{for all } w \in U_w , \qquad (11.26)$$

where $\|\cdot\|$ is some $L_p$-norm.

Find

$$\hat{u} \in U_b = \{u \in L_2[0,T] : |u(t)| \le 1 \quad \text{a.e. on } [0,T]\}$$

such that with $u_o = \Theta$, $w = \Theta$, $z \in C[0,1]$,

$$\|S\hat{u} - z\| \le \|Su - z\| \qquad \text{for all } u \in U_b , \qquad (11.27)$$

where $\|\cdot\|$ is some $L_p$-norm.

PROPOSITION 11.9: If the initial control problem (11.26) has a minimal value different from zero, then the weak bang-bang principle holds. The same is true for (11.27).

In order to obtain a strong bang-bang principle we have to investigate the normality of S.

THEOREM 11.10: If the initial control problem (11.26) has a minimal value different from zero, then the finite bang-bang principle holds. If the boundary control problem (11.27) has a minimal value greater zero, then the countable bang-bang principle holds.

Proof: Let $\ell \in C^* [0,1]$ be given which satisfies the condition (11.10).

For the initial control (11.25) gives

$$S^*\ell(x) = \sum_{k=1}^{\infty} \exp(-\lambda_k T)\, v_k(x)\, \ell(v_k) \tag{11.28}$$

and for the boundary control

$$S^*\ell(t) = \sum_{k=1}^{\infty} \exp(-\lambda_k(T - t))\, \overline{v}_k(1)\, \ell(v_k) . \tag{11.29}$$

The function (11.28) can be extended analytically from the interval [0,1] to an open neighborhood in the complex plane, whereas the one in (11.29) is extendable analytically only from $[0, T - \varepsilon]$, for each $\varepsilon > 0$. Hence both functions either vanish or have at most finitely many zeroes on these intervals. If they vanish identically, then we can conclude $S^*\ell = \Theta$ which is a contradiction. Therefore, the bang-bang principles stated in the theorem hold.

The difference in the statements of Theorem 11.10 for initial and boundary control disappears if we consider the supremum-norm as cost function. The following theorem has been proved by Karafiat [ 4 ]:

THEOREM 11.11: Consider the boundary control problem (11.27) with the maximum-norm on C [0,1]. If the minimal value is greater then zero, then the finite bang-bang principle holds.

In order to solve these problems numerically, we must truncate the series of the solution operators,

$$S_n u = \int_o^T \sum_{k=1}^n \exp(-\lambda_k(T-s))\, u(s)\, ds\ \ \overline{v}_k(1)\, v_k(x) \tag{11.30}$$

or

$$S_n w = \int_o^1 \sum_{k=1}^n \exp(-\lambda_k T)\, v_k(r)\, w(r)\, dr\, v_k(x)\ . \tag{11.31}$$

and consider problem (11.14).

<u>THEOREM 11.12:</u> If the initial control problem (11.26) has a minimal value different from zero, then there is $n_o \in \mathbb{N}$ such that the finite bang-bang principle holds for the truncated problems with $n \geq n_o$:

Find $\hat{w}_n \in U_w$ such that

$$||S_n \hat{w}_n - z|| \leq ||S_n w - z|| \qquad \text{for all } w \in U_w\ .$$

The same statement is true for the boundary control problem.

<u>Proof:</u> According to Theorem 11.7 we only need to show the normality of $S_n$ and the convergence of $S_n$ to S. For this purpose consider the adjoint operator for the initial control problem:

$$S_n^* \ell = \sum_{k=1}^n \exp(-\lambda_k T)\, \ell(v_k)\, v_k(\cdot)\ .$$

The functions $\{v_k\}_{k=1}^n$ are a Haar system, see [ 5 ]. Since by Definition 11. 5 $S_n^* \ell$ does not vanish identically, it has at most n - 1 zeroes. The same conclusion holds for the boundary control problem if one uses that the functions

$$\{\exp(-\lambda_k(\cdot))\}_{k=1}^n$$

are also a Haar system, see [ 6 ].

The operators $S_n$ are compact, because their ranges are finite dimensional. The convergence in the operator norm follows by inspection from (11.30), (11.31) and (11.22).

Hence it could be possible to approximate a boundary control problem by problems for which the finite bang-bang principle holds, although it does not hold for the original problem itself.

## 11.3 NUMERICAL EXAMPLES

We shall make a few remarks on the numerical solution of the boundary control problem (11.27). First we discuss the algorithm presented in Section 7.4 where the switching points are combined by convex combinations. The problem to which the algorithm is applied is given by

$$
\begin{aligned}
&y_t(t,x) = y_{xx}(t,x)\ , && t \in (0,T),\quad x \in (0,1)\ ,\\
&y(0,x) = 0 && x \in [0,1]\ ,\\
&\frac{1}{b}\, y_x(t,1) + y(t,1) = v(t) && t \in (0,T]\ ,\\
&y_x(t,0) = 0\ , && t \in (0,T]\ ,\\
&v'(t) = a(u(t) - v(t)), && t \in [0,T]\ ,\\
&v(0) = 0\ . && 
\end{aligned}
\tag{11.32}
$$

The model has been extended such that the temperature v(t) of the surrounding medium is changed by the fuel rate u(t) which is described by the ordinary differential equation (11.32). The bounds on the control are

$$U = \{u \in L_\infty\,[0,T] : 0 \le u(t) \le 1\}\ .$$

We are approximating the solution operator S by $S_n$ as described in the previous section.

Select a = 25, b = 10, z = 0.2, N = 10.

As cost function we take the $L_1$-norm

$$\int_0^1 |S_n u(x) - z(x)|\,dx\ .$$

In Figure 11.1 we find the computed optimal controls for T = 0.15, 0.2, 0.25.

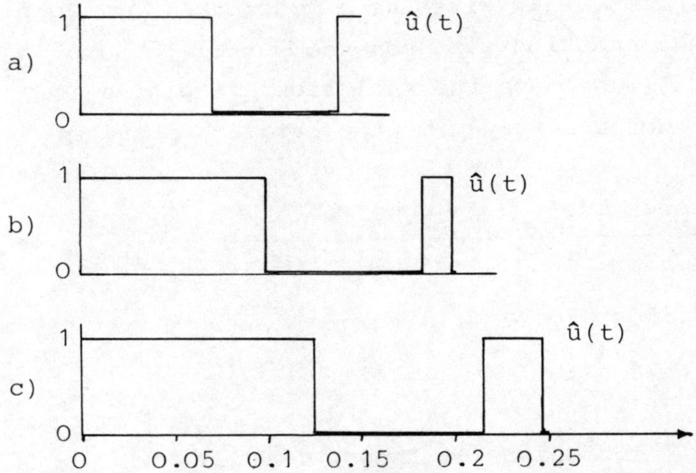


Figure 11.1 Optimal controls

The corresponding final temperature distributions are sketched in Figure 11.2.

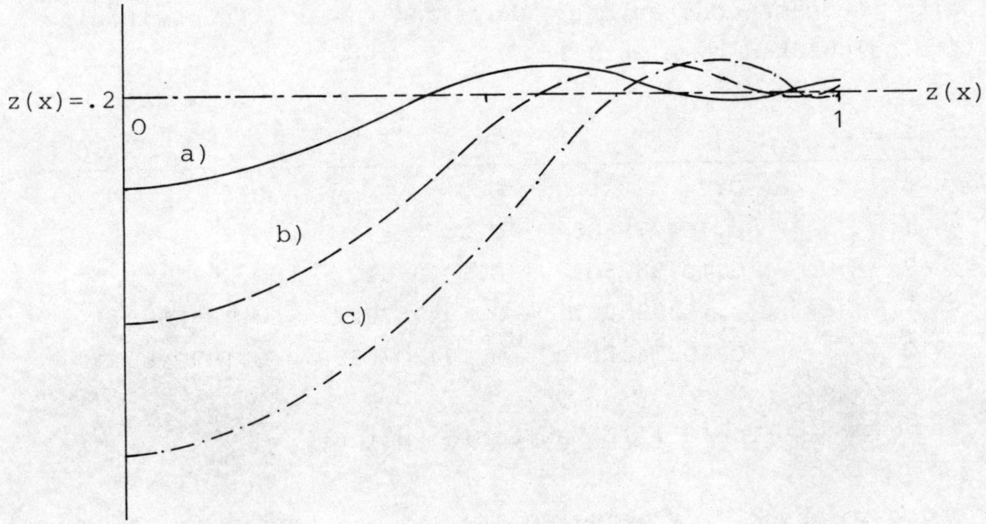


Figure 11.2 Optimal trajectories

We can calculate upper and lower bounds of the optimal value for the truncated problems. This gives us a condition for stopping the algorithm. In connection with some computable error estimates (see [7] ), we have the inclusions displayed in Table 11.1 of the optimal value for the original problem.

| time T | lower bd. | upper bd. |
|---|---|---|
| 0.15 | $6.630 \cdot 10^{-2}$ | $6.642 \cdot 10^{-2}$ |
| 0.20 | $3.650 \cdot 10^{-2}$ | $3.655 \cdot 10^{-2}$ |
| 0.25 | $1.426 \cdot 10^{-2}$ | $1.462 \cdot 10^{-2}$ |

Table 11.1 Inclusions of the optimal value

For the $L_2$-norm in the objective functional

$$\left( \int_0^1 (S_n u(x) - z(x))^2 dx \right)^{1/2}$$

we have the following results:

With the same constants as above and n = 30, the switching points calculated by Algorithm 7B are given in Table 11.2.

| iter. | $t_1$ | $t_2$ | $t_3$ |
|---|---|---|---|
| 0 | 0.2 | 0.2 | 0.2 |
| 1 | 0.104979688 | 0.2 | 0.2 |
| 2 | 0.104982509 | 0.194958202 | 0.2 |
| 3 | 0.103825697 | 0.194957927 | 0.199999811 |
| 10 | 0.103546251 | 0.194307695 | 0.199996500 |

Table 11.2 Switching points

In order to check how close we are to the optimum we compute $\bar{u}_{10}$. The switching points of $\bar{u}_{10}$ however are

$$t_1 = 0.103515711, \quad t_2 = 0.194272243, \quad t_3 = 0.199990653$$

Although 10 iterations were already performed, this indicates that the accuracy of the optimal location of the switching points still can be improved. Using a Newton type iteration where the points $t_1$, $t_2$, $t_3$ are variables starting after iteration 3 we obtain results shown in Table 11.3.

| iter. | $t_1$ | $t_2$ | $t_3$ |
|---|---|---|---|
| 3 | 0.103825697 | 0.194957927 | 0.199999811 |
| 4 | 0.103530683 | 0.194246245 | 0.199995262 |
| 5 | 0.103540201 | 0.194300298 | 0.199989962 |
| 6 | 0.103540272 | 0.194300638 | 0.199990553 |
| 7 | 0.103540272 | 0.194300638 | 0.199990553 |

Table 11.3 Switching points

The switching points of $\bar{u}_7$ coincide with those of $u_7$ such that the switching points of the optimal control are those of $u_7$ in Table 11.3. The inclusion for the optimal value $\rho$ is

$$4.9416 \cdot 10^{-2} \leq \rho \leq 4.9432 \cdot 10^{-2} .$$

Projection methods also have been applied to this problems. The projection of an arbitrary function g into the set given by (11.31) is

$$P_g(t) = \begin{cases} 1 & & g(t) \geq 1 \\ g(t) & \text{for } t \text{ such that} & g(t) \in [0,1] \ , \quad t \in [0,T] \ , \\ 0 & & g(t) \leq 0 \ . \end{cases}$$

However this has the disadvantage that at each iteration controls are computed which are not of bang-bang type. For some numerical results see Denn [ 8 ].

Another way of approximating these problems is to use a Ritz-Galerkin method.

Replace U by

$$U_n = \{n \in L_\infty \, [0,T] : u(t) = \sum_{i=0}^{n} v_i \chi_i(t),$$

$$v_i \in [0,1], \quad i = 0,\ldots,n\}$$

where $\chi_i$ is the characteristic function on the interval

$[t_i, t_{i+1})$ for some partition

$$0 = t_0 < t_1 < \dots < t_{n+1} = T .$$

Then $Su$ for $u \in U_n$ can be written as

$$\overline{S}_n v = \sum_{i=0}^{n} \int_{t_i}^{t_{i+1}} G(T - s, x, 1)\, ds\, v_i = \sum_{i=0}^{n} \gamma_i(x) v_i$$

Corresponding to the norm in which we measure the deviation of the temperature from z, we obtain different optimization problems:

$$L_2\text{-norm:} \quad F(v) = \int_0^1 \Big( \sum_{i=0}^{n} \gamma_i(x)\, v_i - z(x)\Big)^2 dx \ ,$$

$$L_1\text{-norm:} \quad F(v) = \int_0^1 \Big| \sum_{i=0}^{n} \gamma_i(x)\, v_i - z(x)\Big| dx \ ,$$

$$L_\infty\text{-norm:} \quad F(v) = \max_{0 \le x \le 1} \Big| \sum_{i=0}^{n} \gamma_i(x)\, v_i - z(x)\Big| dx \ .$$

In the first case we obtain a quadratic optimization problem, in the second and third cases a finite convex optimization problem with possibly nondifferentiable cost function or semi-infinite linear programming problems. This approach has been used in [9] and [10]. The optimal controls are not bang-bang because the partition was not chosen to coincide with the switching points of the optimal control.

## 11.4 COMMENTS

Section 11.1: The definitions and theorems are an extension of [1].

Section 11.2: Bang-bang principles for linear parabolic control problems are treated in [3], [4],[11] - [15]. For results on controllability we refer to the review article of Russell in [16].

Section 11.3: Numerical results with gradient type or Ritz-Galerkin procedures are obtained in [14], [8] - [10]. Methods based on a variation of the switching point are used in [1], [7], [17] and [18], partially in order to obtain time-optimal controls.

See also the references [19] and [20] for further results.

# 12 Linear hyperbolic control problems

## 12.1 VIBRATING SYSTEMS

Consider the problem of a vibrating system such as a string or a membrane whose motion is controlled at the boundary. Let the deviation of the system from zero at time $t \in (0,T)$ and at point $x \in \Omega$ ($\Omega$ a bounded, open domain in $\mathbb{R}^n$) be denoted by $y(t,x)$. Then the differential equation which describes this motion is

$$y_{tt}(t,x) = \Delta y(t,x), \qquad t \in (0,T), \quad x \in \Omega, \tag{12.1}$$

where $\Delta$ is the Laplacian operator.

At the $C^\infty$-boundary $\partial\Omega$ of $\Omega$ we impose the following condition, $a > 0$,

$$a \frac{\partial y}{\partial \nu} (t,\xi) + y(t,\xi) = g(t,\xi) \qquad t \in (0,T), \quad \xi \in \partial\Omega. \tag{12.2}$$

There are two conditions for the initial state

$$y(0,x) = y_0(x) , \qquad x \in \Omega, \tag{12.3}$$

$$y_t(0,x) = y_1(x) , \qquad x \in \Omega, \tag{12.4}$$

Similarly as for the parabolic case we can distinguish between boundary control $g$ and initial control $y_0$, $y_1$.

Under suitable smoothness assumptions

$$y_0 \in C^2(\overline{\Omega}) , \qquad y_1 \in C^1(\overline{\Omega}) ,$$

$$g_0(t,\xi) = v(t)\, \gamma(\xi) , \qquad v \in C\,[0,T], \ \gamma \in C^2(\overline{\Omega}) ,$$

the system (12.1) - (12.4) can be transformed into an equivalent one with homogeneous boundary conditions, see [ 1 ].

Let $s$ be a solution of the elliptic problem

$$\Delta s(x) = 0 , \qquad x \in \Omega , \tag{12.5}$$

$$a \frac{\partial s}{\partial \nu} (\xi) + s(\xi) = \gamma(\xi) , \qquad \xi \in \partial\Omega \tag{12.6}$$

Then y can be written as

$$y(t,x) = \bar{y}(t,x) + v(t)\, s(x) , \qquad (t,x) \in (0,T) \times \Omega , \qquad (12.7)$$

with $\bar{y}$ defined as the solution of

$$\bar{y}_{tt}(t,x) - \Delta\bar{y}(t,x) = \bar{f}(t,x) \qquad t \in (0,T),\ x \in \Omega , \qquad (12.8)$$

$$a \frac{\partial \bar{y}}{\partial \nu}(t,\xi) + \bar{y}(t,\xi) = 0 \qquad t \in (0,T),\ \xi \in \partial\Omega , \qquad (12.9)$$

$$\bar{y}(0,x) = \bar{y}_0(x), \qquad x \in \Omega, \qquad (12.10)$$

$$\bar{y}_t(0,x) = \bar{y}_1(x), \qquad x \in \Omega, \qquad (12.11)$$

$$\bar{f}(t,x) = f(t,x) - v'(t)\, s(x) \qquad \text{on} \quad (0,T) \times \Omega , \qquad (12.12)$$

$$\bar{y}_0(x) = y_0(x) - v(0)\, s(x) \qquad \text{on} \quad \Omega , \qquad (12.13)$$

$$\bar{y}_1(x) = y_1(x) - v'(0)\, s(x) \qquad \text{on} \quad \Omega , \qquad (12.14)$$

The solution formula for y is given by (see [ 1 ])

$$y(t,x) = v(t)\, s(x) + \sum_{k=1}^{\infty} \Big( \cos \sqrt{\lambda_k}\, t \int_{\Omega} \bar{y}_0(r)\, v_k(r)\, dr$$

$$+ \frac{1}{\sqrt{\lambda_k}} \sin \sqrt{\lambda_k}\, t \int_{\Omega} \bar{y}_1(r)\, v_k(r)\, dr$$

$$+ \frac{1}{\sqrt{\lambda_k}} \int_0^t \sin \sqrt{\lambda_k}\, (t - s) \int_{\Omega} \bar{f}(s,r)\, v_k(r)\, dr ds \Big)\, v_k(x) \qquad (12.15)$$

Formula (12.15) will be specialized for particular problems in the later context.

## 12.2 OPTIMAL CONTROL OF THE VIBRATING STRING

We obtain a model for a vibrating string if we consider the differential equation (12.1) with boundary and initial conditions (12.2) - (12.4) for a one dimensional system.

As the objective of the control problem we choose the energy at a certain time T

$$\frac{1}{2} \int_0^1 (y_x(T,x)^2 + y_t(T,x)^2)\, dx .$$

For the functions defining the initial conditions we require that

$$y_0(0) = y_1(0) = 0 . \tag{12.16}$$

The boundary condition is specified by

$$y(0,t) = 0 , \qquad t \in (0,T) ,$$

$$y(1,t) = v(t) , \qquad t \in (0,T) . \tag{12.17}$$

(12.17) indicates that the string is held fixed at the left side and is moved at the right side according to the control v(t). We are not controlling the movement at the right side directly, but we control the acceleration of the motion. Hence the control set is

$$V = \{v \in C^1 [0,T] : |v''(t)| \le 1 \text{ a.e. in } [0,T],\ v'' \in L_2[0,T],$$

$$v'(0) = 0, \quad v(0) = y_0(1)\} .$$

<u>LEMMA 12.1:</u> The solution of the problem of the vibrating string under the stated assumption is given by

$$y(t,x) = v(t)x + \hat{y}(t,x) \tag{12.18}$$

$$- 2 \sum_{k=1}^{\infty} (-1)^{k+1} (k\pi)^{-2} \int_0^t v''(s) \sin (k\pi (t - s))ds \sin k\pi x,$$

where for $t \in [0,T]$, $x \in [0,1]$

$$\hat{y}(t,x) = \frac{1}{2} (\hat{y}_0(x + t) + \hat{y}_0(x - t) + \int_{x-t}^{x+t} \hat{y}_1(s) ds)$$

with

$$\hat{y}_0(x) = \psi_0(x) - \chi(x)\ y_0(1) ,$$

$$\hat{y}_1(x) = \psi_1(x) - \chi(x)\ v'(0) ,$$

$$\chi(x) = x, \quad \chi(x + 2) = \chi(x), \qquad x \in [-1,1] ,$$

$$\psi_i(x + 2) = \psi_i(x) = \begin{cases} y_i(x) & x \in [0,1] \\ -y_i(-x) & x \in [-1,0] \end{cases} \qquad i = 0,1 .$$

Proof: (12.18) is a rewritten version of (12.15). Under the given boundary conditions, a solution of (12.5) and (12.6) is given by

$$s(x) = x .$$

By inspection, $\hat{y}$ solves

$$\begin{aligned} \hat{y}_{tt}(t,x) &= \hat{y}_{xx}(t,x) , & t \in (0,T), \quad & x \in (0,1) , \\ \hat{y}(0,x) &= y_0(x) - xy_0(1) , & & x \in (0,1) , \\ \hat{y}_t(0,x) &= y_1(x) - xv'(0) , & & x \in (0,1) , \\ \hat{y}(t,0) &= \hat{y}(t,1) = 0 , & & t \in (0,T) . \end{aligned}$$

Hence, the infinite sum in (12.18) represent the inhomogeneity in the differential equation and is identical with the last summand in (12.15).

Instead of the set V we define the following for the control set

$$U = \{u \in L_\infty [0,T] : |u(t)| \le 1 \quad \text{a.e. on } [0,T]\} .$$

For each $u \in U$ there exists with the given initial conditions a unique element $v \in V$ and conversely. Choosing

$$G = L_2 [0,T]^2 ,$$

the cost functional can be written as

$$||Su - z||^2$$

with

$$s(x) = (s_1(x), s_2(x)) ,$$

$$s_i(x) = \frac{1}{2} (\psi_0' (x + T) + (-1)^i \psi_0' (x - T) + \psi_1 (x + T) - (-1)^i \psi_1 (x - T)) \quad x \in [0,1], \quad i = 1,2 .$$

and

$$Su = (S_1u, S_2u) ,$$

$$S_1u(x) = -x + 2 \sum_{k=1}^{\infty} \frac{(-1)^{k+1}}{k} \int_0^T u(t) \cos(k\pi(T-t))dt \sin k\pi x,$$

$$S_2u(x) = T - \int_0^T (T - t)\, u(t)\, dt$$

$$+ 2 \sum_{k=1}^{\infty} \frac{(-1)^{k+1}}{k} \int_0^T u(t) \sin(k\pi(T-t))dt \cos k\pi x \; .$$

Hence the optimal control problem is of the same type as in Section 11.1. However, the questions of controllability depend also on the types of boundary conditions and compatibility conditions and are not discussed here. Normality cannot be expected for these problems, but for the approximate problems with the truncated series this property holds.

There are several examples discussed in [ 2 ]:

Truncation of S and solution by the conditional gradient method with fixed step size rule give for T = 1.5 and

$$y_1(x) = 0 \; , \qquad x \in [0,1] \; ,$$

$$y_0(x) = \sin \pi x \; , \qquad x \in [0,1] \; ,$$

the following switching points for the optimal control $\hat{u}_n$ in Table 12.1.

| n | $t_1$ | $t_2$ |
|---|---|---|
| 5 | 0.117219 | 1.161129 |
| 10 | 0.116951 | 1.161342 |
| 20 | 0.117128 | 1.161956 |
| 40 | 0.117134 | 1.161905 |

Table 12.1 Switching points of $\hat{u}_n$ for T = 1.5

If T is not larger than 2, then we can dispense with the Fourier expansion of the solution and compute the solution via d'Alembert's principle. In [ 3 ] the following lemma is proved.

<u>LEMMA 12.2:</u> If $T \le 2$, then the energy

$$\frac{1}{2} \int_0^1 (y_t(T,x)^2 + y_x(T,x)^2)\, dx$$

of the vibrating string at time T can be expressed as

$$\int_0^T (g(t) - v'(t))^2 \, dt + \int_T^2 g(t)^2 \, dt \,, \tag{12.19}$$

$$g(t) = \begin{cases} \frac{1}{2} (y_1 (1 - t) - y_0' (1 - t)) & t \in [0,1] \\ \frac{1}{2} (-y_1 (t - 1) - y_0' (t - 1) & t \in [1,2]. \end{cases} \tag{12.20}$$

Since the second integrand in (12.19) is constant it can be omitted in the minimization problem:

Find

$$\hat{u} \in U = \{u \in L_\infty [0,T] : |u(t)| \leq 1 \quad \text{a.e. in } [0,T]\} \tag{12.21}$$

such that $||Su - z||$ is minimal with

$$G = L_2 [0,T], \qquad z = g \,,$$

$$Su = \int_0^t u(s)ds \,.$$

It is trivial to see that no bang-bang principle holds for this problem, even if the minimal value is smaller than one. In fact, this problem can be written as a linear optimal control problem with ordinary differential equations and bounds on the controls:

Minimize

$$\frac{1}{2} \int_0^T (g(t) - x(t))^2 \, dt \tag{12.22}$$

subject to

$$\dot{x}(t) = u(t), \qquad t \in [0,T], \qquad x(0) = 0$$

and

$$u \in U \qquad \text{(by (12.21))} \,.$$

A necessary condition for this problem is [ 4 ]:

If $(\hat{u}, \hat{x})$ are the optimal controls and trajectories, then

$$\frac{d}{dt} \hat{x}(t) = \hat{u}(t) \,, \qquad \hat{x}(0) = 0 \,, \tag{12.23}$$

$$\frac{d}{dt} p(t) = -h(t) - \hat{x}(t) \,, \qquad \hat{p}(T) = 0 \,, \tag{12.24}$$

and

$$\hat{u}(t) = -\operatorname{sgn} p(t) , \qquad t \in [0,T] . \tag{12.25}$$

If there is a "singular interval", i.e. $p(t) \equiv 0$ on $[t_o, t_1]$, then the derivative of p also vanishes in this interval and $\hat{x}(t) = h(t)$ on $[t_o, t_1]$. This implies

$$\hat{u}(t) = \frac{d}{dt}\hat{x}(t) = \frac{d}{dt} h(t) \qquad \text{in} \quad [t_o, t_1].$$

In [ 2 ] the conditional gradient method has been applied to this problem with

$$y_1(x) = B \sin \pi q x , \qquad y_o(x) = A \sin \pi q x ,$$

for some constants A, B and q.

Then (12.20) yields

$$g(t) = \begin{cases} \frac{1}{2}\,(B \sin (\pi q(1-t)) - Aq\pi \cos (\pi q(1-)) & t \in [0,1] \\ \frac{1}{2}\,(-B \sin (\pi q(t-1)) - Aq\pi \cos (\pi q(t-1)) & t \in [1,2]. \end{cases}$$

For example, $A = 1$, $B = 0$, $q = 1$ and $T = 0.5$ imply

$$g(t) = -\frac{\pi}{2} \cos (\pi(1-t)) .$$

Then the problem can be interpreted geometrically as follows: Approximate g in the $L_2$-norm by piecewise linear functions, whose slopes have the maximal value 1. Two examples are shown in Figure 12.1.

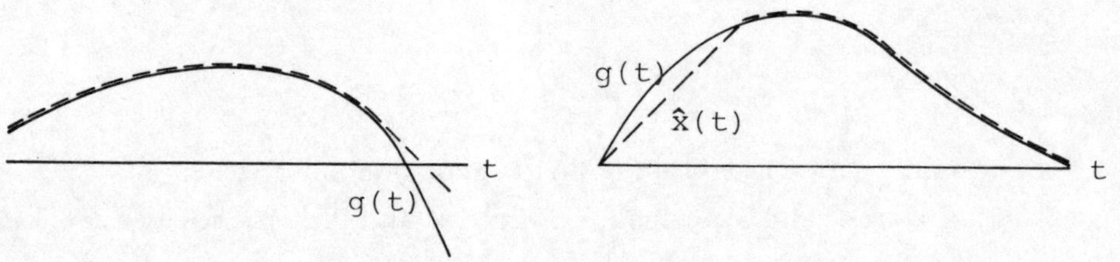


Figure 12.1 Optimal Controls with singular arcs

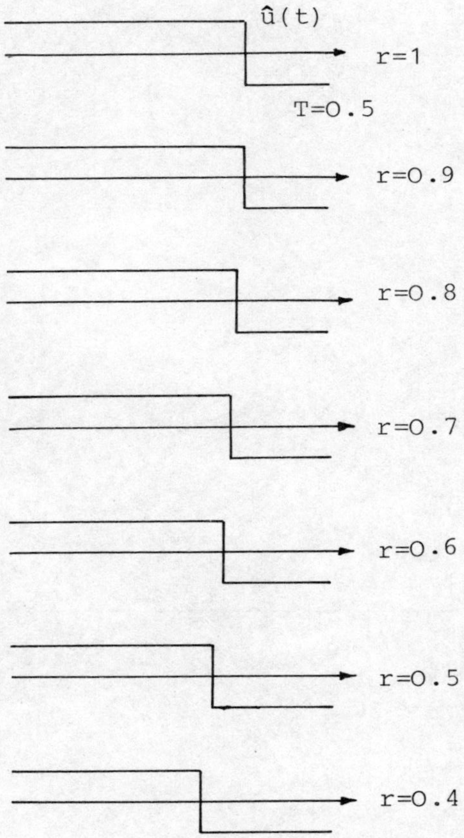


Figure 12.2 Optimal controls depending on parameter q ($y_o(x) = r \sin\pi x$)

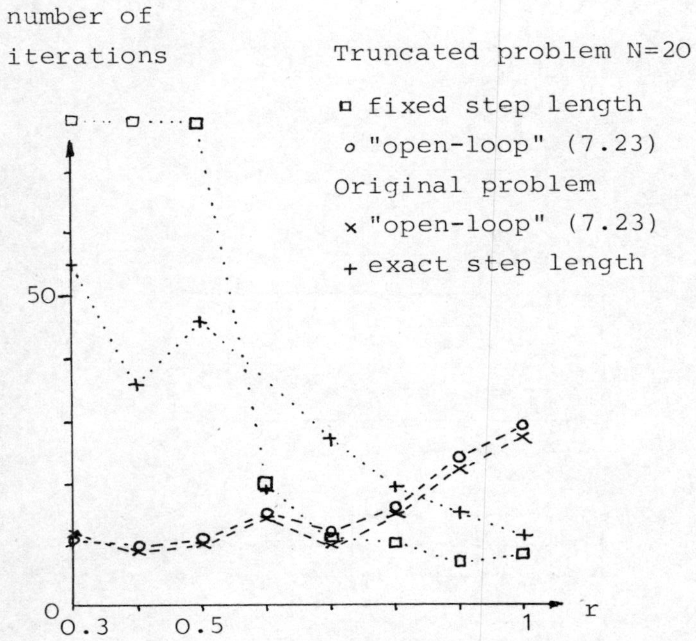


Figure 12.3 Number of iterations of conditional gradient method depending on step length rule and r

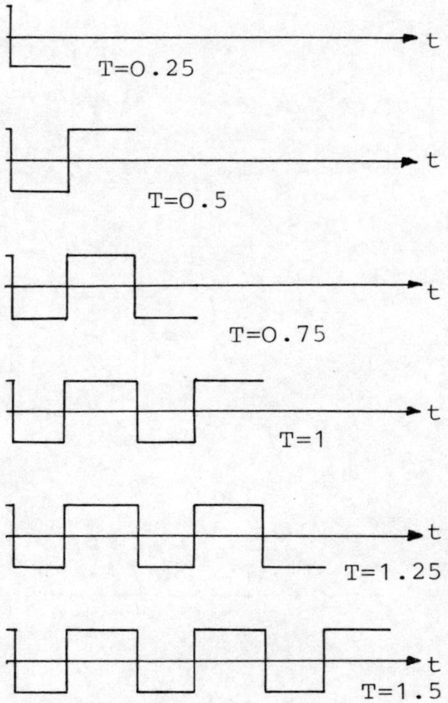


Figure 12.4 Optimal controls depending on final time T

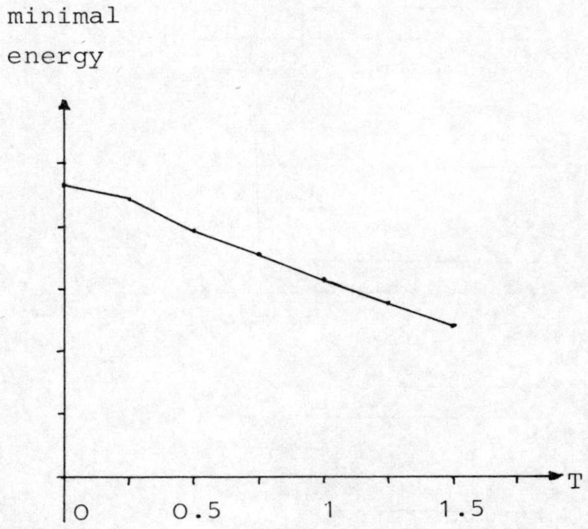


Figure 12.5 Minimal energy depending on final time T

The Figures 12.2 and 12.3 show results obtained with the conditional gradient method using exact and fixed step lengths. Figure 12.5 displays the dependence of the energy on the time interval for $q = 4$, $A = 1$, $B = 0$. In addition Figure 12.4 gives an indication of the growth of the number of switching points.

Let us discuss a few aspects of the class of methods which vary the switching points more or less directly. The control

$$u(t) = \varepsilon(-1)^i, \quad t \in [t_i, t_{i+1}), \quad i = 0,\dots,m-1$$

yields via (12.23) the state

$$x(t) = \varepsilon \left( \sum_{j=1}^{i} (-1)^{j-1} (t_j - t_{j-1}) + (-1)^i (t - t_i) \right)$$

$$t \in [t_i, t_{i+1}) \qquad i = 0,\dots,m-1$$

and the objective is written by

$$\sum_{i=0}^{m-1} \int_{t_i}^{t_{i+1}} \left( \sum_{j=1}^{i} (-1)^{j-1}(t_j - t_{j-1}) + (-1)^i (t - t_i) - g(t)\right)^2 dt .$$

The minimization of this function is a nonlinear finite dimensional minimization problem. The constraints

$$0 = t_0 \le t_1 \le \dots \le t_{m+1} = T$$

can be taken into account by adding a penalty term to the objective. For example, a control with eight jumps was calculated, see Figure 12.6 for $T = 2$, $q = 4$.

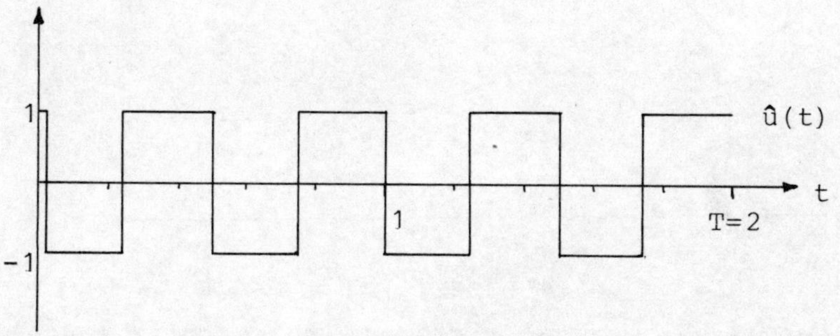


Figure 12.6 Optimal control with 8 switching points

However, as pointed out in [2], if the starting point is not properly chosen, certain difficulties can occur. In contrast to parabolic control problems, the switching points are well distributed over the whole interval and they do not accumulate.

A more direct approach for (12.23) - (12.25) is as follows. If $\hat{u}$ has switching points at $t_1, \ldots, t_m$, then the adjoint function p in (12.24) has m zeroes at $t_1, \ldots, t_m$.

$$p(t) = \int_t^T g(s) + \hat{x}(s))ds = \int_t^T g(s) \quad + \quad \int_0^s u(\sigma)d\sigma ds$$

$$0 = p(t_i) = \int_{t_i}^T g(s)ds$$
$$+ \int_{t_i}^T \left( \sum_{j=1}^{i} (-1)^{j-1} (t_j - t_{j-1}) + (-1)^j (t - t_j)\right)dt ,$$

a system nonlinear equations in $t_1, \ldots, t_m$.

As we have seen for these hyperbolic control problems, a bang-bang principle does not hold. Hence, in the case of singular arcs, discretization type methods are needed. Proceeding as in Section 11.4, a quadratic optimization problem is obtained. A control of this type is displayed in Figure 12.7.

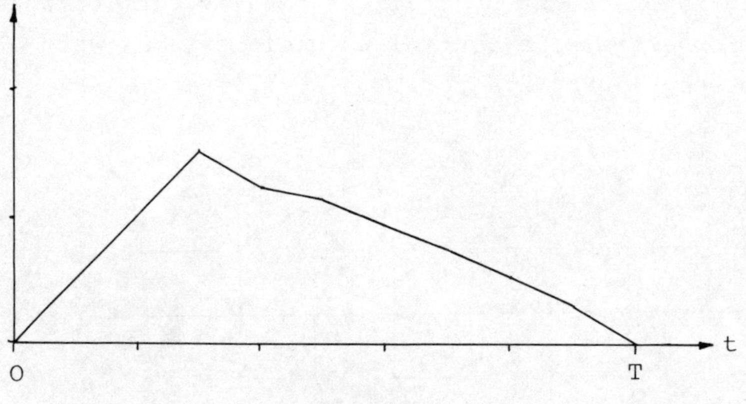


Figure 12.7 Optimal control with singular arc

## 12.3 SOME OTHER HYPERBOLIC CONTROL PROBLEMS

In this section, we briefly mention some other problems which also belong to the class of hyperbolic control problems.

The vibrating beam

$$\begin{aligned}
y_{tt}(t,x) &= y_{xxxx}(t,x) & t \in (0,T),\quad x \in (0,1),\\
y(0,x) &= y_0(x),\quad y_t(0,x) = y_1(x), & x \in (0,1),\\
y(t,0) &= y_{xx}(t,0) = 0, & t \in (0,T),\\
y(1,t) &= v_1(t), & t \in (0,T),\\
y_{xx}(1,t) &= v_2(t), & t \in (0,T).
\end{aligned}$$

The boundary controls lie in

$$v_1 \in V_1 = \{v \in V_2 : \; v(0) = 0\} \quad ,$$

$$v_2 \in V_2 = \{v \in C^1 \,[0,T] : v'' \in L_2 \,[0,T]\}.$$

We assume the compatibility conditions,

$$y_0(0) = y_0(1) = y_0''(0) = y_0''(1) = y_1(0) = 0 \; .$$

The energy of the beam at time T is expressed as

$$\frac{1}{2} \int_0^1 (y_t(T,x)^2 + y_{xx}(T,x)^2)\, dx \; .$$

Let us set

$$v_1(0) = v_2(0) = v_1'(0) = v_2'(0) = 0 \; .$$

There are three ways to control the beam:

a) amplitude control: $v_2 \equiv 0$, $v_1$ variable

b) moment control: $v_1 \equiv 0$, $v_2$ variable

c) vector-valued control: $v_1$, $v_2$ variable

Results for

$$y_0(x) = \sin \pi x, \quad y_1(x) \equiv 0$$

using these different controls are shown in Figure 12.8.

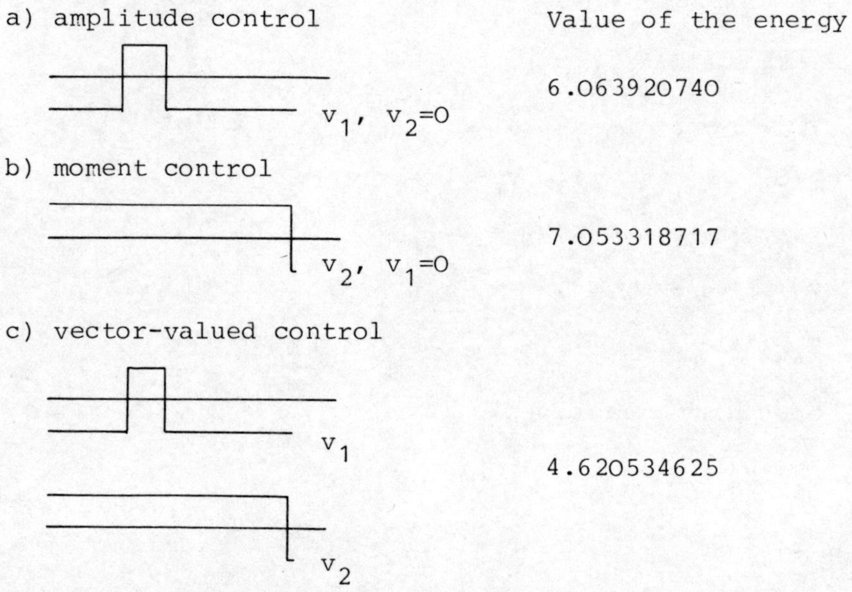


Figure 12.8 Optimal controls of a vibrating beam

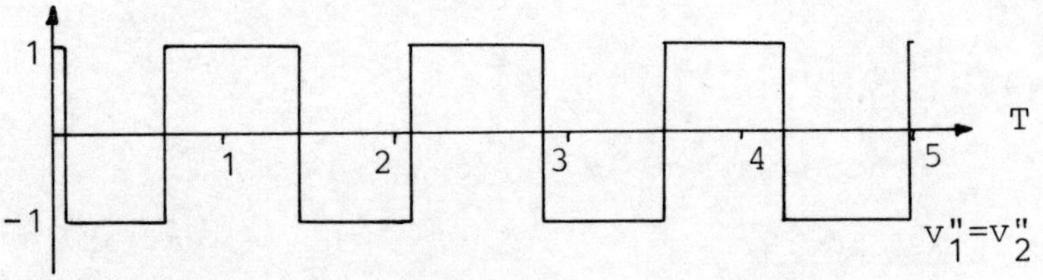


Figure 12.9 Optimal controls of a vibrating membrane

A two dimensional hyperbolic system is given by the vibrating membrane: $\Omega : (0,1) \times (0,1)$.

$$\begin{aligned}
y_{tt}(t,x) &= \Delta y(t,x), & t \in (0,T), \quad x \in \Omega, \\
y(0,x) &= y_0(x), \;\; y_t(0,x) = y_1(x), & x \in \Omega, \\
y(\xi,t) &= r_1(\xi)v_1(t) + r_2(\xi)v_2(t), & t \in (0,T), \quad \xi \in \partial\Omega, \\
r_1(x_1,x_2) &= (1 - x_1)\,(1 - x_2), & \\
r_2(x_1,x_2) &= x_1 x_2 . &
\end{aligned}$$

The edges at (1,0) and (0,1) are held fixed while the edges at (1,1) and (0,0) are moved. With $v(0) = v'(0) = 0$ and

$$y_0(x) = \sin \pi x_1 \;\; \sin \pi x_2 , \qquad y_1(x) = 0$$

the initial energy

$$\frac{1}{2} \int_\Omega (y_t(T,x)^2 + y_{x_1}(T,x)^2 + y_{x_2}(T,x)^2)\, dx$$

is 2.4674011. This is reduced to 0.451993145 allowing 5 time units and the controls are shown in Fig. 12.9.

## 12.4 COMMENTS

The authors would like to thank W. Eichenauer for the numerical results in this chapter.

Section 12.1: For a rigorous treatment of hyperbolic equations we refer e.g. to the book Triebel [1].

Section 12.2: A discussion of the vibrating string problem for $T \leq 2$ is in [3] where also singular optimal controls are mentioned. Controllability for linear hyperbolic systems is reviewed in [6] with a list of references. Further numerical results are published in [2] and [5]. For a discretization approach with splines and estimates corresponding to the smoothness of the optimal control see [7]. Time optimal control problems are treated in [8] and [9]. Since the switching points of bang-bang controls are almost equally distributed over the time interval in contrast to the accumulating effect for parabolic problems, methods where the switching points are varied are quite successful for problems described in this chapter.

Section 12.3: The vibrating beam is treated numerically in [2], [5], [10] and [11]. Vibrating membranes are investigated in [5].

# 13 Minimum-fuel orbital rendezvous

## 13.1 INTRODUCTION

This chapter describes a problem of determining optimal trajectories for interplanetary orbital rendezvous missions [ 1, 2 ]. A space vehicle is to be transferred by the action of a suitable velocity control $u(t) \in \mathbb{R}^3$, for t in a certain time interval, from a prescribed initial orbit to a prescribed final orbit at an unspecified final time. The control is obtained as an "impulsive approximation"

$$u(t) = \sum_{i=1}^{N} c_i \, \delta_i \, (t - t_i) , \tag{13.1}$$

where the parameters $c_i \in \mathbb{R}^3$ and $t_i \in \mathbb{R}$, $i = 1,\ldots,N$, denote the velocity increment $c_i$ which is added at time $t_i$ to the current velocity of the vehicle. N is the number of impulses. The state vector

$$x(t) = (r(t), v(t)) \in \mathbb{R}^6 \tag{13.2}$$

is composed of the position r(t) and velocity of the vehicle at time t, relative to some coordinate system. In this formulation, we consider the dynamics of "two body motion" in a central force field, where the bodies are the earth and the spacecraft, each object represented by a point mass.

The general form of the two body equation is

$$\dot{x}(t) = \begin{pmatrix} \dot{r}(t) \\ \dot{v}(t) \end{pmatrix} = \begin{pmatrix} v(t) \\ - \mu r(t) \, \|r(t)\|^{-3} \end{pmatrix} + \begin{pmatrix} 0 \\ u(t) \end{pmatrix} \tag{13.3}$$

where $\mu$ is a known gravitational constant. The total fuel consumption corresponding to the control (13.1) is

$$\sum_{i=1}^{N} \|c_i\|$$

and the variables are the parameters $t_1,\ldots,t_N$, $c_1,\ldots,c_N$ such that

$$\underline{t}_i \le t_i \le \overline{t}_i , \qquad c_i \le \|c_i\| \le \overline{c}_i , \qquad i = 1,\ldots,N$$

with fixed bounds on the switching times and velocity increments.

## 13.2 IMPULSIVE CONTROL MODEL

A simpler formulation of the latter minimum-fuel control problem, which also leads itself to a solution by methods of nonlinear programming, is to decompose the impulsive controlled motion of the state $x(t)$ during $[t_o, t_N]$ as a sequence of equivalent free (uncontrolled) motions during the time subintervals $[t_i, t_{i+1}]$, $i = 0,\ldots, N-1$, with different initial conditions (Fig. 13.1):

$$\dot{x}_i(t) = \begin{pmatrix} \dot{r}_i(t) \\ \dot{v}_i(t) \end{pmatrix} = \begin{pmatrix} v_i(t) \\ -\mu r_i(t)\,\|r_i(t)\|^{-3} \end{pmatrix} \quad \begin{matrix} t \in [t_i, t_{i+1}], \\ i = 0,\ldots,N-1, \end{matrix}$$

$$x_i(t_i) = \begin{pmatrix} r_i(t_i) \\ v_i(t_i) \end{pmatrix} = \begin{pmatrix} r_{i-1}(t_i) \\ v_{i-1}(t_i) + c_i \end{pmatrix} \quad i = 1,\ldots,N . \tag{13.4}$$

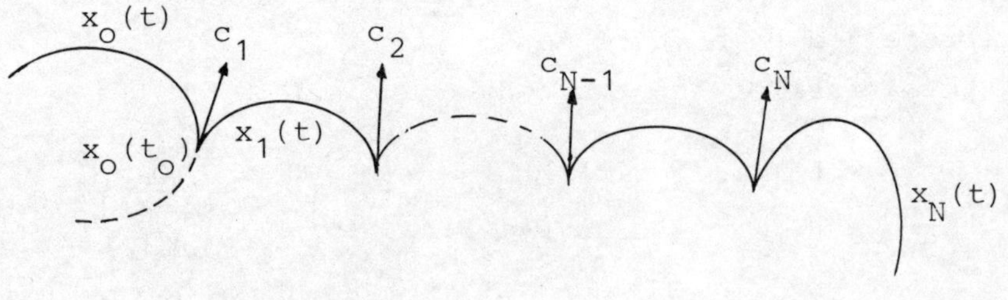


Figure 13.1 Sketch of problem decomposition

Initially, the space vehicle lies in a prescribed elliptical orbit which is described as a function of certain variables called Kepler elements: a, the semimajor axis; e, the eccentricity; i, the inclination; $\Omega$, the angle of the ascending mode; $\omega$, the argument of perigee; $M(\tau)$, the eccentric anomaly at time $\tau$. In the degenerate case of circular motion, treated in Section 13.3, the Kepler elements are replaced by another appropriate set of orbital elements. Thus, the initial state is determined by

$$x_o(t) = \psi_o(t) = \psi(a, e, i, \Omega, \omega, E(\tau), t) .$$

Similarly, the spacecraft is to be steered onto an orbit, also determined by another given set of orbital elements,

$$\psi_N(t) = \psi(\bar{a}, \bar{e}, \bar{i}, \bar{\Omega}, \bar{\omega}, \bar{E}(\bar{\tau}), t) .$$

Hence, we have the following set of boundary conditions for $x_i(t)$:

$$\begin{aligned} x_1(t_1) &= \psi_o(t_1) + \begin{pmatrix} 0 \\ c_1 \end{pmatrix} , \\ x_i(t_i) &= x_{i-1}(t_i) + \begin{pmatrix} 0 \\ c_i \end{pmatrix} \qquad i = 2,\ldots,N , \\ x_{N-1}(t_N) &= \psi_N(t_N) . \end{aligned} \tag{13.5}$$

The equation of two body motion has been extensively studied in celestial mechanics. The solutions of this equation, as predicted by Kepler's law, are conic sections. However, the precise form of the solution depends on the control; that is, whether circular, elliptical, parabolic, or hyperbolic motion is obtained. Convenient methods of computing the solution have been developed by various investigators, e.g. the universal variable approach given by Goodyear [ 3 ], and using this approach (13.4) can be solved in terms of the initial data:

$$x_i(t) = \rho_i(t) = G(t - t_{i-1}, x_{i-1}(t_i) + \begin{pmatrix} 0 \\ c_i \end{pmatrix}) ,$$

for some nonlinear functions $\rho_i$, $i = 1,\ldots,N$. Hence, we can regard the original optimization problem as a nonlinear programming problem:

$$\text{minimize} \quad F(p) = \sum_{i=1}^{N} \|c_i\| \qquad (13.6)$$

with

$$p(t_1, t_N, c_1, \ldots, c_N) \in \mathbb{R}^{4N}$$

subject to

$$\underline{t} \le t_1 \le t_2 \le \ldots \le t_N \le \overline{t} \ , \quad t_i \in \mathbb{R} \qquad (13.7)$$

$$\underline{c}_i \le c_i \le \overline{c}_i \ , \quad c_i \in \mathbb{R}^3 \ , \quad i = 1, \ldots, N \qquad (13.8)$$

$$\xi_1 = \psi_0(t_1) \ ,$$

$$\xi_{i+1} = G(t_{i+1} - t_i, \xi_i + \binom{0}{c_i}) \ , \quad i = 1, \ldots, N-1 \ ,$$

$$\xi_N = \psi(t_N) \ , \qquad (13.9)$$

where $\xi_i \in \mathbb{R}^6$ represents $x_{i-1}(t_i)$ as a dummy variable that can be eliminated in (13.9) such that we obtain N - 1 nonlinear equations in the variables $t_1, \ldots, t_N, c_1, \ldots, c_N$.

## 13.3 NUMERICAL EXAMPLES

The preceding optimization problem in (13.6) - (13.9) was solved using a projection-restoration algorithm with the BFGS variable metric update and step length determined by gradient and function evaluation. Solution of the two body equations and evaluation of the state partial derivatives used in the gradient algorithm was obtained using a computer program developed by Goodyear [ 4 ]. The computation was carried out on a Telefunken TR 440 computer in double precision (92 bit wordlength). Inequality constraints were incorporated in the algorithm using an active set strategy.

Problem 1. Comet Rendezvous [ 5 ]

For scientific investigations of the comet Encke as it approaches our sun, the flyby velocity of a spacecraft as it approaches the comet, should be small. A final orbit $\psi_N$ was chosen which met this latter requirement, and two impulses ($N = 2$) were allowed to transfer the spacecraft from the initial orbit to its final orbit. Starting with an approximation of the switching times and velocity increments such that (13.7) and (13.8) are not active, the data of Table 13.1 were assumed:

| Initial Orbit | Final Orbit |
|---|---|
| $a = 15 \cdot 10^8$ km | $\bar{a} = 3.3225\ 10^8$ km |
| $e = 0.0167$ | $\bar{e} = 0.8462$ |
| $i = 0^\circ$ | $\bar{i} = 12.35^\circ$ |
| $\omega = 101.2208^\circ$ | $\bar{\omega} = 185.2^\circ$ |
| $\Omega = 0^\circ$ | $\bar{\Omega} = 334\ 72$ |
| $M(\tau) = -101.2208^\circ$ | $M(\bar{\tau}) = 0^\circ$ |
| $\tau$ = Sept. 23, 1980 | $\bar{\tau}$ = Dec. 2, 1980 |

Table 13.1 Input Orbital Data

In Table 13.1 we specify the mean anomaly $M(\tau)$ rather than the eccentric anomaly. Using an initial approximation of

$$t_1 = 0 \text{ days} \qquad c_1 = (-5.9, -5.9, 0) \text{ km/sec}$$
$$t_2 = 100 \text{ days} \qquad c_2 = (-11, -13, 4) \text{ km/sec}$$

where the times are relative to Sept. 23, 1980, the converged values of the control parameters,

$$\hat{t}_1 = -24.75 \text{ days} \qquad \hat{c}_1 = (-3.005, -8.421, 4.112) \text{ km/sec}$$
$$\hat{t}_2 = 81.16 \text{ days} \qquad \hat{c}_2 = (-1.102, -6.536, -2.416) \text{ km/sec} ,$$

corresponding to a launch date of Aug. 29, 1980 and a flight time of 106 days. The injection velocity (increment) is 9.87 km/sec and the velocity (increment) to attain flyby is 7.05 km/sec. Approximately 20 sec. CPU time was required to obtain convergence on the Telefunken TR 440. The norm of the constraint violations (13.9) was less than $10^{-6}$ and the norm of the gradient of the Lagrangian was less than $10^{-3}$ at termination of the algorithm.

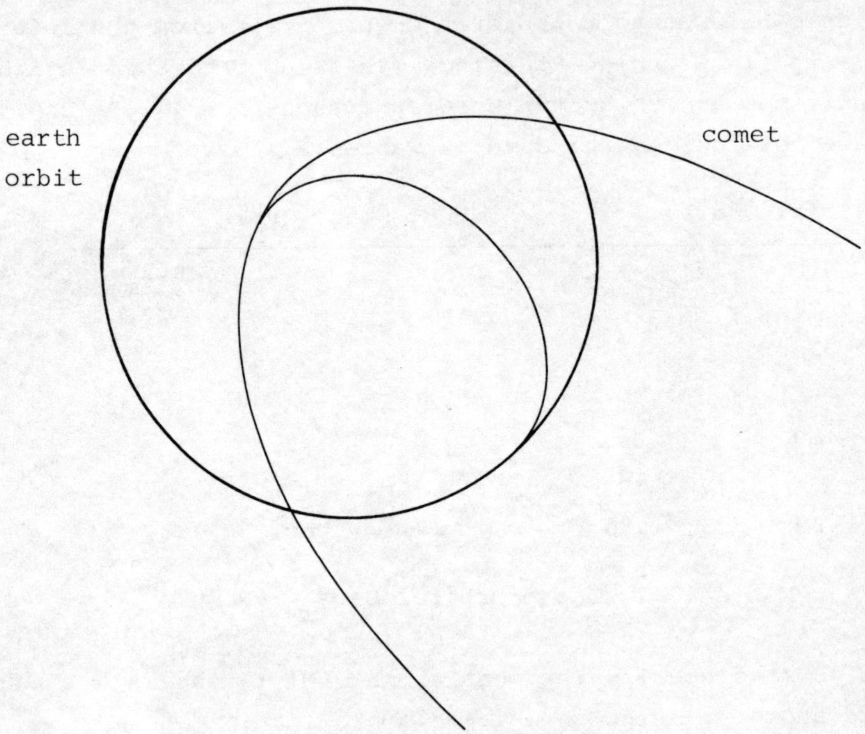


Figure 13.2  Rendezvous comet Encke

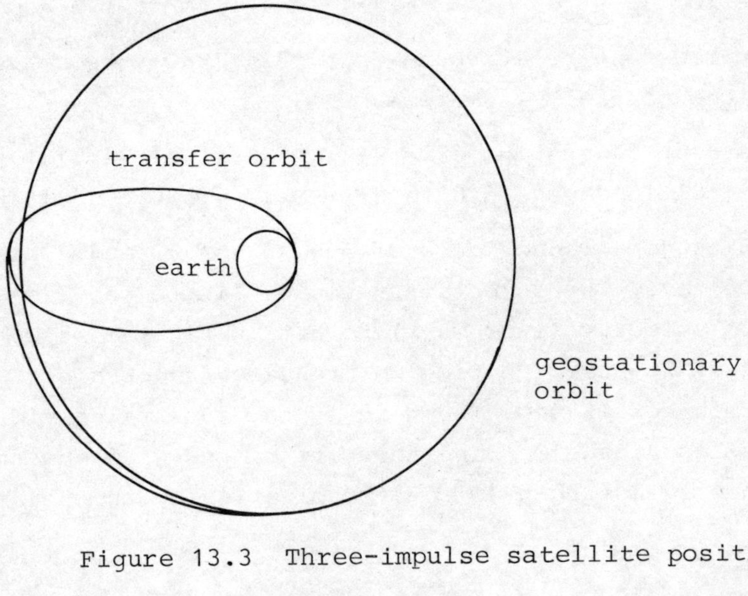


Figure 13.3  Three-impulse satellite positioning

| | Initial values | Converged values Problem 1 | Problem 2 | Problem 3 | Problem 4* | Problem 5* |
|---|---|---|---|---|---|---|
| $t_1$ | 0 hr | - 0.242 | - 0.143 | 0.017 | 0.0 | 0.0 |
| | 0 m/sec | 125.5 | 72.3 | 1.94 | 0.0 | 0.0 |
| $c_1$ | - 1420 | - 1452.7 | - 1443.8 | - 1435.6 | - 1424.5 | - 1424.5 |
| | 0 | - 125.4 | - 128.9 | - 111.9 | - 124.6 | - 124.6 |
| $t_2$ | 12 hr | 12.84 | 11.68 | 11.33 | 9.54 | 8.0 |
| | 0 m/sec | - 26.4 | - 0.04 | 17.6 | - 0.76 | 14.4 |
| $c_2$ | - 50 | - 29.6 | - 25.6 | - 29.3 | - 13.8 | - 50.8 |
| | 0 | 10.6 | 4.58 | 16.7 | 16.2 | 14.9 |
| $t_3$ | 24 hr | | 23.2 | 22.8 | 27.3 | 25.0 |
| | 0 m/sec | | - 0.40 | - 1.06 | - 45.8 | - 51.7 |
| $c_3$ | - 20 | | - 7.8 | - 14.5 | - 70.7 | - 30.2 |
| | 0 | | - 2.26 | - 8.5 | - 2.46 | - 3.67 |
| | | 1463.5 | 1451.4 | 1440.0 | 1430.0 | 1430.0 |
| | | 41.1 | 25.9 | 38.1 | 21.3 | 54.8 |
| | | | 8.1 | 16.8 | 84.2 | 60.0 |
| Total cost m/sec | | 1504.6 | 1485.4 | 1494.9 | 1535.5 | 1544.8 |

*For these cases, the first parameter vector $(t_1, c_1)$ is fixed (not optimized).

Table 13.2 Satellite positioning (rendezvous)

| Problem | Cycles* | Function evaluations | | Computer time** | Parameters | Constraints | |
|---|---|---|---|---|---|---|---|
| | | Restoration | Projection | (in seconds) | n | p | q |
| Earth-Mars | 5 | 25 | 22 | 7 | 8 | 7 | |
| 1 | 6 | 25 | 19 | 6 | 8 | 6 | |
| 2 | 11 | 40 | 41 | 19 | 12 | 6 | |
| 3 | 15 | 61 | 66 | 31 | 12 | 6 | 1 |
| 4 | 15 | 79 | 100 | 27 | 8 | 6 | |
| 5 | 25 | 113 | 137 | 44 | 8 | 6 | 2 |

*Stopping tolerances: $\varepsilon_1 = 10^{-3}$, $\varepsilon_2 = 10^{-6}$.

**Telefunken TR 440 Computer (double precision).

Table 13.3 Convergence properties of the projection-restoration algorithm

Problem 2. Geostationary Satellite Positioning

A communication satellite is to be transfered from a given elliptical orbit to a geostationary (circular) orbit by impulse control. It is assumed that N = 3 impulses are employed to attain rendezvous.

| Initial orbit | Final orbit |
|---|---|
| $a = 25078$ km | $r(t_A) = (-42164.22,0,0)$ km |
| $e = 0.736047$ | $v(t_A) = (0, -3074.65,0)$ km/sec |
| $i = 5^\circ$ | |
| $\omega = 180^\circ$ | |
| $\Omega = 180^\circ$ | |
| $E(t_A) = 180^\circ$ | |
| $t_A$ = time of apogee passage in the transfer (initial) orbit | |

Table 13.4 Input Orbital Data

Position is relative to an earth centered coordinate system with the x and y axes in the plane of the equator. In place of the Kepler elements, the position on a circular orbit is defined by the coordinates of state and velocity at time $t_A$. For this problem, the following different cases are studied. Unless otherwise stated, the constraints (13.7), (13.8) were not active during the iteration due to the initial starting point. The computation time was 33 sec. on the TR 440.

Case 1: N = 2.

This case is used as a basis for comparison. The required apogee thrust $\|c_1\|$, shown in Table 13.2, is 1463.5 m/sec with the final correction $\|c_2\| = 411$ m/sec occuring 13.08 hours after insertion into the primary orbit.

Case 2: N = 3.

In comparison to case 1, less fuel is required for the mission due to the additional corrective impulse in the secondary orbit. A fuel savings of about 20 m/sec suggests the advisability of designing a three impulse maneuver. The final converged trajectory is displayed in Fig. 13.3.

Case 3: N = 3, $||c_1|| \leq 1440$ m/sec.

The bound on the apogee thrust $||c_1||$ reflects either a fixed apogee motor filling, e.g. a solid fuel rocket which burns for a fixed duration, or a limit of tank capacity. Even with this restriction, the maneuver is still superior to the two impulse case.

Case 4: N = 3, $t_1$ = 0 hrs, $c_1$ = (0, -1424.5, - 124.6) m/sec.

This case can also be viewed as an orbit transfer problem with two impulses (N = 2), however with a different initial orbit, that which is attained after insertion. The motivation for this case is to show the effect of errors, both in magnitude and direction of the apogee motor.

Case 5: N = 3, $t_1$ = 0 hrs, $c_1$ = (0, -1424.5, -124.6) m/sec,
$t_2$ 8 hrs, $||c_3|| \leq 60$ m/sec.

In this case the time $t_2$ of the first correction is constrained to occur at least 8 hrs. after insertion with the primary orbit, for example, to allow sufficient time for orbit measurement by tracking stations. The third impulse $c_3$ has a bound of 60 m/sec, defining a possible system limitation.

## 13.4 COMMENTS

Section 13.1: The problem of determining fuel-optimal impulsive trajectories has an extensive literature dating to the work of Lawden in the early fifties [ 6 ]. Subsequent extensions of Lawden's results using the "primer vector" approach were obtained by Lion and Handelsman [ 7 ] and Jezewski and Rozendaal [ 8 ]. A comprehensive survey of the impulsive trajectory problem appears in [ 9 ].

Section 13.2: The dynamics of two body motion using Kepler elements is treated in most texts on orbital mechanics, see for example [ 10]. Gradients of the objective and constraints with respect to the parameters $t_i$, $c_i$ can be easily computed using the chain rule for differentiation [ 1 ]. The required state partial derivatives are supplied by the two body solution method [ 4 ,10 ].

Section 13.3: The problem formulation, orbital data, and solution algorithm described in Problem 2 were utilized in the development of a positioning program for the German-French satellite Symphonie, which was successfully launched from Cape Canaveral several years ago.

# 14 Optimal control of dynamic queueing systems

## 14.1 INTRODUCTION

This section describes an application of optimal control to stochastic service systems, which are modeled as finite capacity Markov chains, with time varying arrival and service rate [1,2] A special case of the general model derived below is a multi-programmed computer system (Figure 14.1) where the computer is treated as a single channel server with mean service rate $\mu$ per pass through the central processing unit. With probability q, a job completing a pass will require additional computation and will be sent to the front of the job queue for further processing. Arrivals and computation completions occur according to a Poisson process. The arrival rate of jobs fluctuates with peaks during mid-morning and mid-afternoon each day. The cost of operating the system consists of the direct cost of providing a level of service $\mu$ as well as the indirect costs due to a backlog of jobs. An optimal service rate minimizes the sum of the costs.

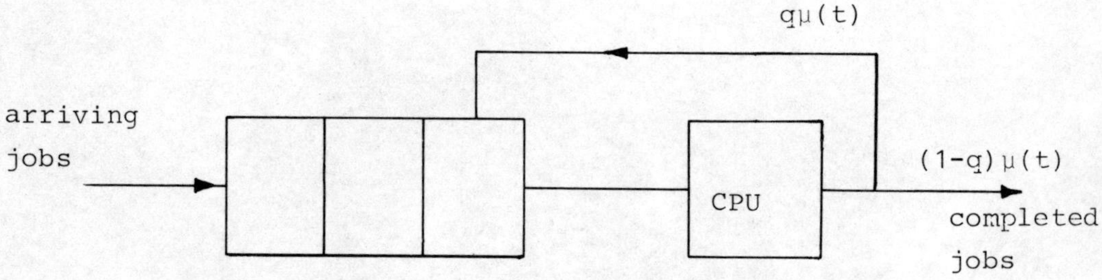


Figure 14.1 Multiprogrammed computer system

## 14.2 BIRTH-DEATH MODEL

Let the birth rate of a population of size n, $n = 0,1,\ldots,N$, be denoted by $\lambda_n(t)$, $t \in [t_o,T]$, $T < \infty$, and

$$\overline{\lambda}(t) = (\lambda_o(t), \ldots, \lambda_{N-1}(t)) .$$

Similarly, the death rate of a population of size n is denoted by $\mu_n(t)$, and

$$\overline{\mu}(t) = (\mu_1(t), \ldots, \mu_N(t)).$$

The following standard assumptions, where h is the length of the time interval, are imposed on the system:

(A1) The probability of a birth in $[t,t+h]$ is $h\lambda_n(t) + o(h)$.

(A2) The probability of a death in $[t,t+h]$ is $h\mu_n(t) + o(h)$ .

(A3) The probability of two or more births (or deaths) in $[t,t+h]$ is $o(h)$ where $\lim_{h \to o} o(h)/h = 0$.

Let the probability that the system has n customers be denoted by $x_n(t)$ and

$$\overline{x}(t) = (x_o(t), \ldots, x_N(t)) . \tag{14.1}$$

The capacity of the system will be assumed finite, $N < \infty$. Under the preceding assumptions $x_n(t)$ satisfies the Chapman-Kolmogorov equations, [3]:

$$\dot{x}_n(t) = \lambda_{n-1}(t)\, x_{n-1}(t) - [\lambda_n(t) + \mu_n(t)]\, x_n(t) + \mu_{n+1}(t)\, x_{n+1}(t) \quad 1 \le n \le N - 1 , \tag{14.2}$$

$$\dot{x}_o(t) = - \lambda_o(t)\, x_o + \mu_1(t)\, x_1(t) , \tag{14.3}$$

$$\dot{x}_N(t) = \lambda_{N-1}(t)\, x_{N-1}(t) - \mu_{N-1}(t)\, x_N(t) , \tag{14.4}$$

with initial conditions $(x_o(t_o), \ldots, x_N(t_o))$.

In vector form, (14.2) - (14.4) can be written

$$\dot{\bar{x}} = \begin{pmatrix} -\lambda_0 & \mu_1 & & & 0 \\ \lambda_0 & -\lambda_1+\mu_1 & \ddots & & \\ & \lambda_1 & \ddots & \ddots & \\ & & \ddots & \ddots & \mu_N \\ 0 & & & \lambda_{N-1} & -\mu_N \end{pmatrix} \bar{x} \ , \quad \bar{x}(t_o) = \bar{x}^o \qquad (14.5)$$

The objective

$$F(u) = \int_{t_o}^{T} L(t, \bar{x}(t), u(t))dt$$

represents, for example, the expected cost incurred during the time interval $[t_o,T]$. The control function, identified with either the service or arrival rates $u(t) = \bar{\lambda}(t)$ or $u(t) = \bar{\mu}(t)$ is selected from a bounded set

$$U = \{u \in L_\infty^N [t_o,T] : a \le u^{(i)}(t) \le b, \ i = 1,\ldots,N\},$$
$$\text{for some } a, b \ge 0 \ . \qquad (14.6)$$

Since $x_n(t)$ represents a probability, the following point constraints must be imposed:

$$\sum_{n=0}^{N} x_n(t) = 1 \ , \quad x_n(t) \ge 0 \ , \quad t \in [t_o,T], \quad n = 0,\ldots,N \ .$$

In this particular model it can be shown [1] that if the initial state $\bar{x}(t_o)$ satisfies the probability constraints, then so does $x_n(t)$, $t \in [t_o,T]$. That is, if $\sum_{n=0}^{N} x_n(t_o) = 1$, $x_n(t_o) \ge 0$, $n = 0,\ldots,N$ and $\bar{x}(t)$ satisfies (14.5) and (14.6), then

$$\sum_{n=0}^{N} x_n(t) = 1, \quad x_n(t) \ge 0, \quad n = 0,\ldots,N, \quad t \in [t_o,T].$$

## 14.3 SPECIFIC QUEUEING MODELS

The literature on optimization of queueing systems generally considers steady state solutions of the preceding model to obtain analytical properties of the probabilities. An advantage of the optimal control formulation is the wide latitude in modeling including the ability to choose alternative control structures.

Example 1. A single server queueing system (M/M/1/N). In this case we choose the same service and arrival rate at each "state".

$$\mu_n(t) = \nu(t) \qquad \lambda_{n-1}(t) = \lambda(t), \qquad n = 1,\ldots,N .$$

Example 2. A multiple server queueing system (M/M/c/N). If

$$\mu_n(t) = \begin{cases} n\,\nu(t) & 1 \le n \le c \\ c\,\nu(t) & c \le n \le N, \end{cases}$$

then the control $\mu$ is also scalar valued. Alternatively, let

$$\mu_n(t) = \begin{cases} \nu_n(t) & 1 \le n \le c \\ \nu_c(t) & c \le n \le N, \end{cases}$$

and the control becomes a vector.

Example 3. Bulk service ($M/M^Y/1/N$). Service is performed in batches assuming that k customers are present in the system. Otherwise all customers are served in one batch. The model is as follows:

$$\dot{x}_o(t) = -\lambda_o(t)\,x_o(t) + \sum_{i=1}^{k} \mu_i(t)\,x_i(t),$$

$$\dot{x}_n(t) = \lambda_{n-1}(t)\,x_{n-1}(t) - (\lambda_n(t) + \mu_n(t))\,x_n(t) + \\ + \mu_{n+k}(t)\,x_{n+k}(t), \quad 1 \le n \le N-1,$$

$$\dot{x}_N(t) = \lambda_{N-1}(t)\,x_{N-1}(t) - \mu_N(t)\,x_N(t).$$

Although Example 3 involves no increase in the dimensionality of the state, $\bar{x}(t)$, the modeling of Erlangian, priority, series and cyclic queues can lead to large numbers of state equations.

## 14.4 NUMERICAL EXAMPLES

Results of computational experiments are presented in this section using the general queueing model (14.5). An algorithm was based on a gradient projection method with Euler discretization of the state equations.

Problem 1. Single Server Queue

In this example we study the behavior of an M/M/1/N queue and the efficiency of the algorithm as the limit on system size is increased. The problem is to minimize

$$F(\mu) = \int_{t_0}^{T} \left[ \sum_{n=1}^{N} (150 + 50\,n)\, x_n(t) + 5\mu^2(t) \right] dt$$

subject to (14.5) with

$$\mu_n(t) = \mu(t), \quad x_{n-1}(t) = \lambda(t), \quad n = 1,\ldots,N,$$

$$\overline{x}(t_0) = (1,0,\ldots,0), \quad \lambda(t) = \sin t + 1, \quad t_0 = 8, \quad T = 9.6,$$

$$\mu \in U = \{u \in L_\infty [t_0,T] \mid 0 \le \mu(1) \le 2\} .$$

The results of this experiment are summarized in Table 14.1 where $x^*$ and $\mu^*$ indicate the converged values of the control and corresponding state, respectively.

| N | CPU Time (sec) | $F(\mu^*)$ | $\max_{t_0 \le t \le T} \lvert x_N^*(t) \rvert$ |
|---|---|---|---|
| 3 | 0.53 | 231.591 | 0.115 |
| 6 | 0.78 | 235.821 | $0.618 \times 10^{-2}$ |
| 9 | 1.02 | 235.890 | $0.113 \times 10^{-3}$ |
| 12 | 1.33 | 235.890 | $0.869 \times 10^{-6}$ |
| 15 | 1.59 | 235.890 | $0.323 \times 10^{-8}$ |
| 18 | 1.82 | 235.890 | $0.648 \times 10^{-11}$ |

Table 14.1 System Size Comparison

In the preceding problem the discretization employed 129 grid points. Computer storage requirements were (4N+52) K bytes. For each N, the algorithm required seven iterations for convergence defined by

$$\|\nabla F(\hat{\mu})\| \le 10^{-4} .$$

The results of Table 14.1 suggest several conclusions. First, if only the system capacity N is varied, the algorithm is very stable with respect to the computational effort. The cost converges at N = 15 to a stable value. Although not indicated, the control function also converges. Indeed, for N = 9, the control is accurate to three digits of the value achieved at N = 15. Therefore, even though the system capacity may be very large, solving a much smaller problem should yield satisfactory approximation of the optimal control. A least squares fit of the data obtains

$$\text{CPU time} = .0844\ N + .302 \quad \text{seconds}$$

indicating that computational times are reasonable, even for N = 1000. Thus core storage, not execution time, is the limiting factor for the application of the technique; e.g. for N = 110, nearly 500K bytes storage is required, however, 10 seconds CPU time is needed to obtain convergence.

One approach for alleviating the storage problem is to employ a continuous version of the algorithm utilizing higher order integration schemes. For a given accuracy, higher order integrators require fewer grid points than Euler's method. However, the additional programming needed to represent the function and to employ a compatible adjoint must be taken into consideration.

<u>Problem 2.</u> Comparing Scalar and Vector Service

This problem compares the computational effort associated with scalar and vector controls in the queueing system model. An M/M/1/3 queue is described by birth-death equations (14.5) with $\lambda_n(t) = \lambda(t)$, $n = 0,\ldots,N-1$ and objective

$$F(\mu) = \int_{t_o}^{T} \sum_{n=1}^{N} x_n(t)\ (r_n + 5\mu_n^2(t))\ dt$$

to be minimized by choice of the service rate $\mu$, where

$$r_1 = 250,\quad r_2 = 300,\quad r_3 = 350 \qquad \lambda(t) = 1 + \sin t \qquad t_o = 8$$

$$r_4 = 400,\quad r_5 = 450,\quad r_6 = 500 \qquad \bar{x}(t_o) = (0,1,0,\ldots 0)$$

$$t_1 = 9.6$$

$$\bar{\mu} \in U = \{\bar{\mu} \in L_\infty\ [t_o,T] \mid\ 0 \le \mu_n(t) \le 2,\ \ n = 1,2,3\}$$

Table 14.2 indicates that the total number of iterations required for solving the scalar control problem remains constant for larger n; however, the computational effort for the vector case drops sharply. Other experiments have shown that computational effort is also sensitive to changes in the initial state, time horizon, and system equation parameters.

| control type | system size, N | min. cost | total iterations | time per iteration (sec) |
|---|---|---|---|---|
| scalar | 3 | 327.7286 | 15 | .062 |
| vector | 3 | 327.3789 | 26 | .100 |
| scalar | 6 | 338.1309 | 16 | .086 |
| vector | 6 | 337.7679 | 7 | .187 |

Table 14.2 Scalar versus vector controls

Problem 3. Multiprogrammed Computer System

The last problem concerns the multiprogrammed computer system described in 14.1. In this example the arrival rate is specified by

$$\lambda(t) = 1 + 0.75 \cos \frac{2\pi}{7} (t - 9)$$

reflecting a peak demand at 9:00 AM and 4:00 PM. The problem is to minimize the cost

$$F(\mu) = \int_{t_o}^{T} [\sum_{n=1}^{N} r_n x_n(t) + R\mu^2(t)]\, dt ,$$

where the second term represents the direct cost of providing the level of service $\mu(t)$ per unit time, e.g. rental costs of equipment, personnel wages, maintenance, etc. The first term of the cost integrand is the expected indirect cost due to a backlog of jobs, and the variable cost of operating the computer (varying in that the cost is incurred only if computation is taking place). Examples of indirect costs are the cost of providing storage for waiting jobs, the cost of idle time for customers waiting for completed jobs, and for commercial facilities, and the cost of lost potential customers due to the present backlog.

For this example it is assumed that

$$N = 3, \quad r_1 = 25, \quad r_2 = 30, \quad r_3 = 40,$$

where costs are expressed in hundreds of dollars. The minimum allowable service rate is zero and the maximum is assumed to be 2 jobs/hour. The system begins at 7:00 AM with no jobs.

The converged optimal control for the case where R = 2.5 and q = 0 is shown in Figure 14.2. The minimum value of cost is \$26.64K. The optimal control for the case where R = 3 and g = 0.5 is similar, with a minimum cost of \$37.79K. Figure 14.2 shows the optimal service rate, with a lag of nearly one hour, is similar in shape to the arrival rate. This suggests the following easily implemented control policy

$$\mu(t) = a\lambda(t-b)$$

where a and b are appropriate scalar parameters.

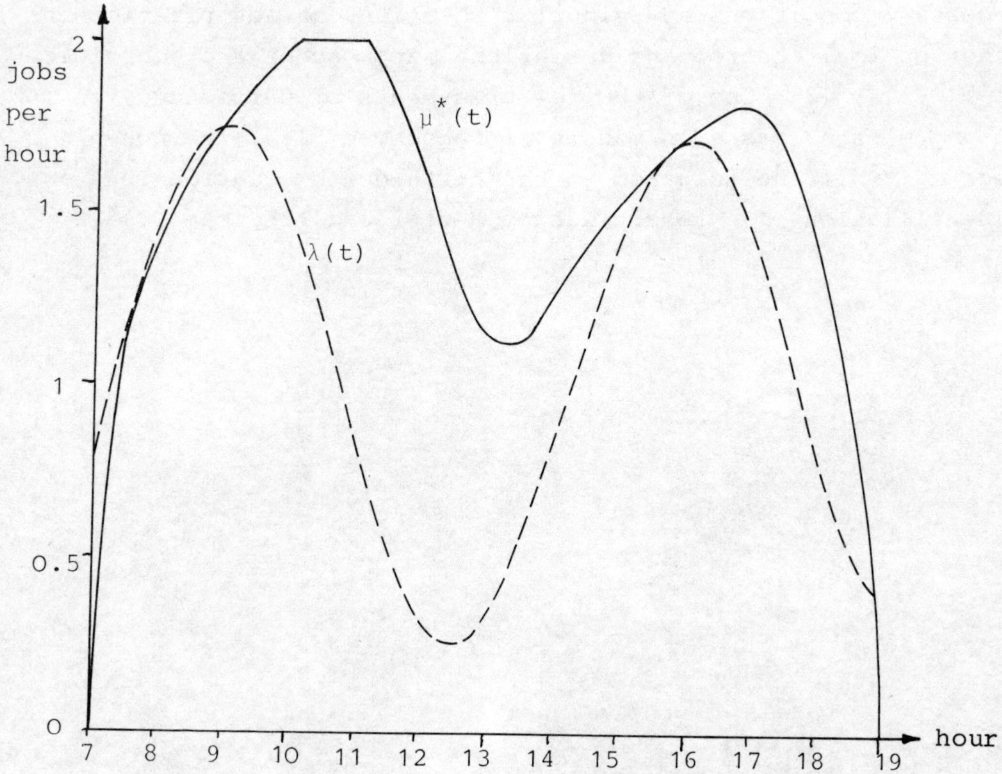


Figure 14.2 Optimal Service Rate

## 14.5 COMMENTS AND REFERENCES

Section 14.1: Until the mid sixties, queuing theory was mainly concerned with descriptive aspects of stochastic service systems such as probability distributions and resulting statistics for steady-state systems. See [4] for a recent survey of optimization in queuing systems, especially for problems with unbounded horizon. A basic text on queuing systems is Gross and Harris [3].

Section 14.2: Miller [5] treats the optimization of bounded horizon, time-invariant queuing systems and cost structures using a finite set of admissible control values for each time point. Related studies have been carried out by Emmons [6] and Martin-Lof [7]. One of the first optimal control models is due to Klimov [8], who obtained necessary conditions based on the Maximum Principle of Pontryagin. Subsequently, Man [9] examined necessary conditions and singular solution characterizations. Other optimal control and variational approaches are contained in [10, 11, 12]. One of the few treatments of sufficiency is in an unpublished Master's thesis of Lansdowne [13]. The non-negativity of the solution to the birth-death equation is a generalization of a result due to Reuter and Lederman [14].

# 15 Optimal planning in integrated production – inventory systems

## 15.1 INTRODUCTION

In this application we consider a deterministic model of a production system for manufacturing products of known demand [ 1, 2]. The products consume available resources, known as "factors of production" - labor, capital and R & D (research and development) engineering. The input-output relationship is given by an assumed or empirically verified production function. Decisions are to be made concerning the factors of production and the product mix for given objectives of the management.

A typical example of this type of problem is the model due to Holt,Modigliani, Muth and Simon [ 3 ], the HMMS model. It determines the production rate and the workforce level for a given demand while minimizing the cost of inputs and inventory. However, HMMS and other related models do not include the possibility of increasing the production rate by R & D engineering activities of the firm during the planning period. Engineering may include research in product and process engineering, industrial engineering, equipment and labor training activities. In the following section we describe a dynamic model of production-inventory systems which includes research and development as an integral part of planning.

## 15.2 PRODUCTION FUNCTION MODEL

A production function is a relationship between the production and the factors of production. Most empirical studies made in the area of production functions [4 , 5 ] can be classified according to the following two types: a linear Leontief type input-output function

$$y(t) = a_1 L(t) + a_2 K(t) \tag{15.1}$$

and a nonlinear Cobb-Douglas type function

$$y(t) = L(t)^{\alpha} K(t)^{\beta} \tag{15.2}$$

where L(t) and K(t) are inputs in labor and capital, respectively, at time t.

Mansfield, et.al. [ 6 ] established the following Cobb-Douglas type production function for planning R & D expenditures

$$y(t) = Ae^{at} \int_{0}^{t} e^{-\lambda s} r(s)ds \; L(t)^{\alpha} K(t)^{1-\alpha} ,$$

where r(s) is the R & D expenditure at time s. This latter function was then expressed by Lele and O'Leary [ 7 ] as

$$y(t) = L(t)^{\alpha} K(t)^{\beta} E(t)^{\gamma} , \tag{15.3}$$

where E(t) is the cumulative engineering expenditure at time t. Thus, it is assumed in (15.3) that labor and capital are consumed as products are manufactured. However, engineering R & D contributes cumulatively to production over the planning period.

The production function (15.3) will be incorporated in an optimal control model for which the following assumptions hold:

(1) The system is able to overcome all delays in processing of the product, and hence, there are no in-process goods.
(2) No defective goods are manufactured.
(3) The required inputs are delivered at the proper time and place.
(4) Inventory is not damaged or stolen, and does not become obsolate.

We define the states I(t) and E(t), the inventory and cumulative engineering, and the controls L(t), K(t) and R(t) which are the rates of labor, capital, and engineering, respectively. Denoting by D(t) the rate of demand, assumed to be known for $t \in [O,T]$, then we have the following state equations

$$\dot{I}(t) = L(t)^{\alpha} K(t)^{\beta} E(t)^{\gamma} - D(t), \quad I(O) = I_0 , \tag{15.4}$$

$$\dot{E}(t) = R(t) , \quad E(O) = E_0 . \tag{15.5}$$

The parameters $\alpha, \beta, \gamma$ are empirically determined technological coefficients of the production function.

An optimization problem is formulated in terms of the minimization of a quadratic objective with fixed cost coefficients for labor ($c_1$), capital ($c_2$), engineering ($c_3$), inventory ($c_4$), terminal inventory ($c_5$):

$$F(u) = \int_0^T (c_1 L(t) + c_2 K(t) + c_3 R(t) + c_4 I(t)^2)dt + c_5 I(T)^2 \qquad (15.6)$$

over all L,K,R satisfying (15.4) - (15.5) and upper and lower bounds

$$\begin{aligned} \underline{L} &\le L(t) \le \overline{L} \\ \underline{K} &\le K(t) \le \overline{K} \qquad \text{for all } t \in [0,T] . \\ \underline{R} &\le R(t) \le \overline{R} \end{aligned} \qquad (15.7)$$

The objective contains linear terms representing the cost due to inputs, and quadratic penalties for a nonzero inventory or backlog (negative inventory).

## 15.3 NUMERICAL EXAMPLES

The optimal control problem formulated in Section 15.2 was discretized using a predictor-corrector method with a mesh size of 40 equally spaced points to obtain a discrete optimal control problem. For this discretized problem the gradient projection method was used for the optimization. Typically, iterations on the IBM 370/165 with double precision wordlength required about 35 direction updates (~ 2 sec. CPU time) to reach convergence determined by $\|\nabla F(L,K,R)\| \le 10^{-5}$.

Data used in analyzing the model is displayed in Table 15.1. Fig. 15.1 shows converged levels of labor, capital, engineering rate and cumulative engineering during the planning period for a typical set of data. The control is bang-bang with no singular arcs. Inventory, I(t), is rapidly driven to zero within the first half of the total interval, and then backlog accumulates. Although not studied in this example, the backlog could be decreased by appropriate selection of the penalty constant, $c_5$.

The model was solved for three different sets of labor and capital costs: (1) $c_1 = 2.$, $c_2 = 5.$, (2) $c_1 = 2.5$, $c_2 = 4.5$, and (3) $c_1 = 3.5$, $c_2 = 3.5$. As the cost of capital (equipment) decreases more capital was used for manufacture of the products.

The cost of labor and capital equipment can be reasonably estimated in any manufacturing environment once the process has been established. However, considerable uncertainty exists in the cost of engineering and inventory. Therefore, the model was solved for different engineering and inventory costs. The final engineering, terminal inventory, and product cost are displayed in Fig.15.2 as a function of inventory costs for different costs of engineering input. Figure 15.2a shows that terminal backlog decreases initially and then increases slightly with inventory costs. Terminal inventory was found to be relatively insensitive to changes in the costs of engineering. We also remark that the backlog increases slightly as the cost of engineering is increased implying that incurring additional backlog is cheaper than more production.

The objective function assumes that the cost of inventory and backlog are equal. However, the cost of backlog may be lost sales which are more expensive than the cost of inventory. Such a modification in the objective can be accomodated by the gradient algorithm, assuming that the differentiability assumptions are satisfied, resulting in an inventory position instead of backlog.

Figure 15.2b indicates that terminal engineering initially increases as the cost of inventory increases and then slightly decreases as a function of the cost of inventory. Similarly the cost of product increases initially and then decreases slightly as a function of inventory. The phenomenon can be explained the following way. As the inventory cost is increased the backlog is decreased by more production during which more inputs including engineering input are consumed, the result of which is reflected by the increased final engineering level and the increased total cost of products in Figs. 15.2b and 15.2c, respectively. The observation is reasonable because the cost of product also includes an increased cost of engineering.

No penalties were assigned for changes in the level of inputs; in practice, however, there is a cost associated with changes in hiring, training, and lay-offs. Such input derivative costs can be easily accomodated in the formulation without significant modification to the algorithm. Similarly, changes in the system constraints can be incorporated. For example, Mansfield [ 6 ] has proposed that productivity of engineering input decreases over the planning period, by inclusion of a suitable discount factor in the production function.

<u>Initial States</u>

| | |
|---|---|
| Inventory | $I(0) = 50$ |
| Engineering | $E(0) = 50$ |

<u>Control Bounds</u>

| | | |
|---|---|---|
| Labor | $\underline{L} = 50$ | $\overline{L} = 65$ |
| Capital | $\underline{K} = 45$ | $\overline{K} = 60$ |
| Engrg. rate | $\underline{R} = 20$ | $\overline{R} = 25$ |

<u>Technological Coefficients</u>

$\alpha = .5$

$\beta = .3$

$\gamma = .2$

<u>Demand (Discretization from 0 to 2 into 40 subdivisions)</u>

100,100,90,70,100,100,80,60,75,90,100,80,50,100,
90,110,70,80,60,100,95,70,85,65,95,80,100,70,
90,85,65,70,90,100,85,100,100,80

Table 15.1 Input Data

| $c_1$ | $c_2$ | $c_3$ | $c_4$ | $c_5$ | Term. Inv. $I(T)$ | Term. Engr. $E(T)$ | Product Cost $\int_0^T (c_1L(t)+c_2K(t)+c_3R(t))dt$ | Inv. Cost $\int_0^T c_4I^2(t)dt$ |
|---|---|---|---|---|---|---|---|---|
| 3 | 4 | 8 | 1 | 0 | -25.4 | 56.7 | 1065 | 1081 |
| | | | 2 | | -23.9 | 58.9 | 1097 | 2135 |
| | | | 3 | | -23.1 | 59.5 | 1107 | 3181 |
| | | | 5 | | -21.5 | 59.7 | 1119 | 5252 |
| | | | 10 | | -16.9 | 59.5 | 1144 | 10200 |
| | | | 15 | | -18.4 | 59.5 | 1135 | 25400 |
| | | | 25 | | -19.8 | 58.5 | 1113 | 25485 |
| 3 | 4 | 15 | 1 | 0 | -25.7 | 54.0 | 1322 | 1085 |
| | | | 2 | | -24.3 | 55.6 | 1363 | 2145 |
| | | | 3 | | -23.2 | 58.0 | 1403 | 3187 |
| | | | 5 | | -21.6 | 59.1 | 1430 | 5257 |
| | | | 10 | | -16.9 | 59.0 | 1455 | 10205 |
| | | | 15 | | -18.4 | 59.5 | 1453 | 15405 |
| | | | 25 | | -19.8 | 58.5 | 1524 | 25484 |
| 3 | 4 | 25 | 1 | 0 | 25.5 | 54.0 | 1723 | 1084 |
| | | | 2 | | -25.8 | 54.0 | 1739 | 2147 |
| | | | 3 | | -23.7 | 54.0 | 1745 | 3203 |
| | | | 5 | | -21.8 | 57.7 | 1846 | 5267 |
| | | | 10 | | -16.9 | 58.1 | 1885 | 10204 |
| | | | 15 | | -18.4 | 59.0 | 1896 | 15411 |
| | | | 25 | | -19.9 | 58.1 | 1859 | 25493 |

Table 15.2 Converged Costs

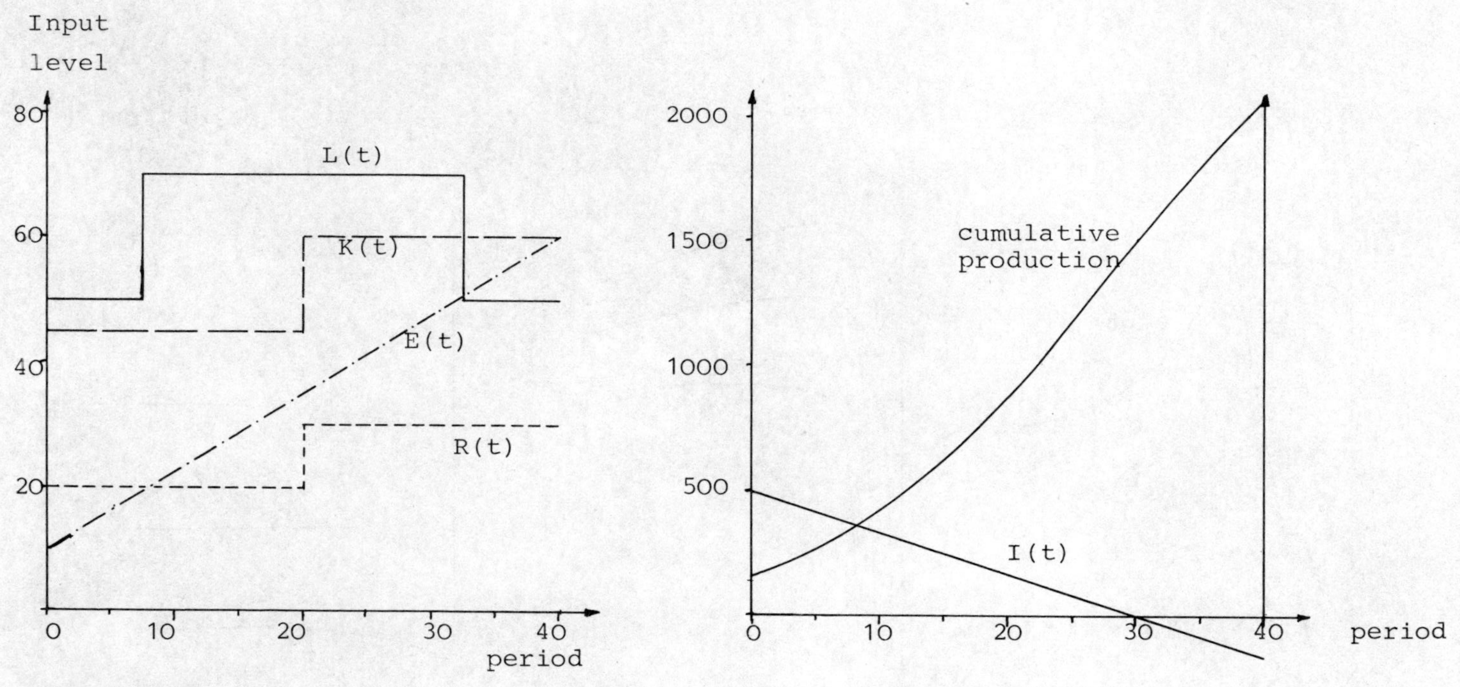


Figure 15.1  Converged levels

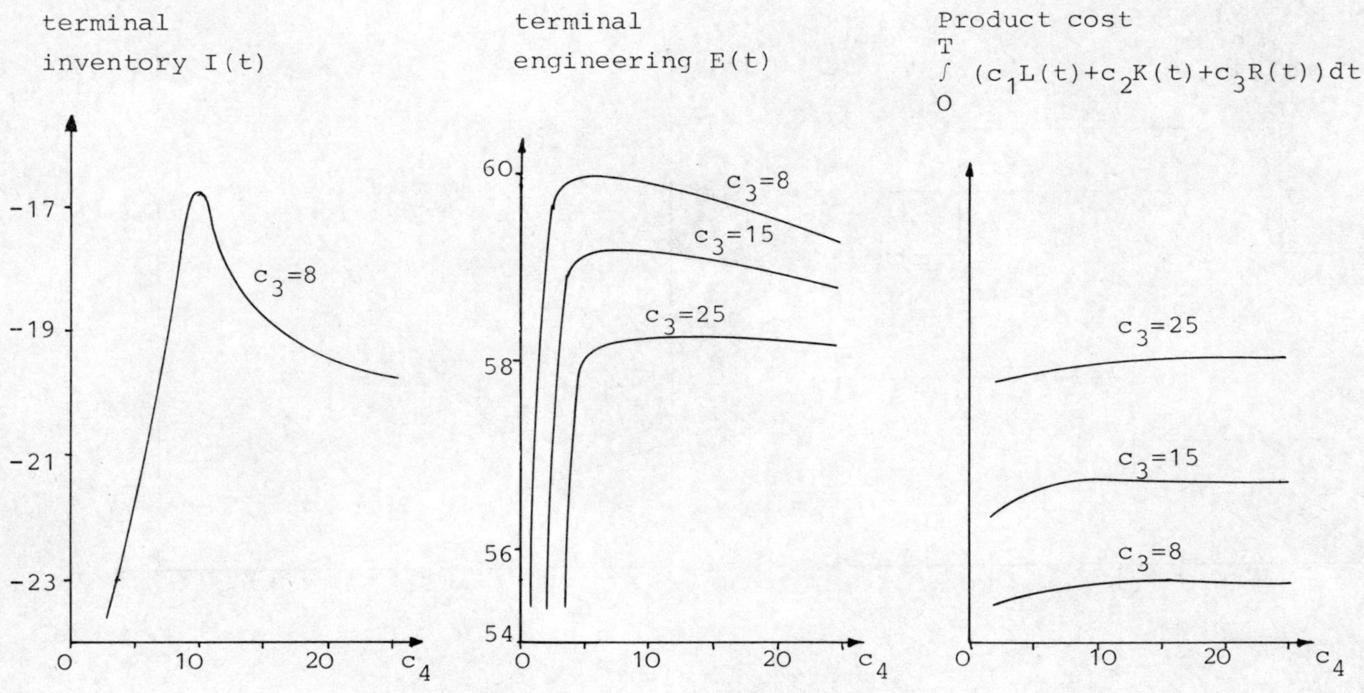


Figure 15.2 Dependence on inventory costs

## 15.4 COMMENTS

Section 15.1: Production functions have been used extensively by economists in the theory of the firm. Applications have been presented by Leontief [ 8 ] in planning and forecasting, Solow [ 9 ] in growth theory, Smith [10] in production planning, Griliches [11] in agricultural production, and Arrow and Kurz [12] in investment and returns.

Section 15.2: Optimal control models in production planning have also been treated by [13-19 ]. See [ 2 ] for a related problem of R & D production-inventory control using a Leontief type production function.

Section 15.3: A comparison of optimal control and linear programming models for production planning appears in [20].

# 16 Optimal evaluation of manual lifting tasks

## 16.1 INTRODUCTION

In this chapter we derive an optimization model for use in the design and evaluation of manual lifting tasks. Most analyses of lifting problems have been concerned with the experimental approach in which methods are used to fit observed data. Such models usually express the physiological cost in terms of some basic task, human, and work place variables.

A second approach for analyzing lifting tasks uses biomechanical models to predict directly the forces involved in a given task. It is this latter approach that we implement in the following section. The fundamental assumption motivating such models is that an individual will optimize his performance consistent with task and work place constraints. Input data for the model, which can be obtained without extensive experimentation, includes

(1) anthropometric characteristics of the person who performs the task,
(2) physical characteristics of the object to be lifted,
(3) initial and final positions of the object, and
(4) the task performance time.

## 16.2 DYNAMIC MODELS OF THE LIFTING TASK

The human body when viewed in the sagittal plane can be considered as five rigid links (Fig. 16.1). These links are of the same length as their corresponding human segments, and possess the same mass and moment of inertia as their human counterpart. Thus, any movement or configuration of the body can be described in terms of the angular positions of these five geometrical links. For the purpose of this study, the link from the shoulder to the hip (spinal column) is considered as one rigid link, i.e., spinal flexion is not considered in the model. It is also assumed that the head and neck are included in the mass of the back. The hands are also assumed to be part of the elbow to wrist structure; that is, in lifting a load, the hands are assumed to have no relative motion with respect to the forearms. In terms of these five links, principles of mechanics are applied below to describe the motion of the body while performing lifting tasks.

The class of motions to be considered by the model is restricted by the following assumptions: (a) the worker does not walk with the load, and (b) the worker remains stationary and lifts the object between two preselected points in the task space. The object, therefore, must be within arm's reach at the beginning of the task and at the end of the task. All movement during the lift is symmetrical, i.e., each arm moves in unison with the other arm (as also do the legs). The ankle is considered fixed on the floor, ignoring the joint at the ball of the foot. Finally, the motions considered during the lift are those motions occuring only in the sagittal plane. For example, body rotation from side to side, i.e., motion that occurs in different planes, is not considered feasible by the model.

To describe the motion of a worker performing a lifting task, a two-dimensional Cartesian coordinate system is assumed, with the origin situated at the ankle joint. The acceleration for an intermediate link about a point (such as the wrist rotating about the elbow) during any instant t of its motion can be written as:

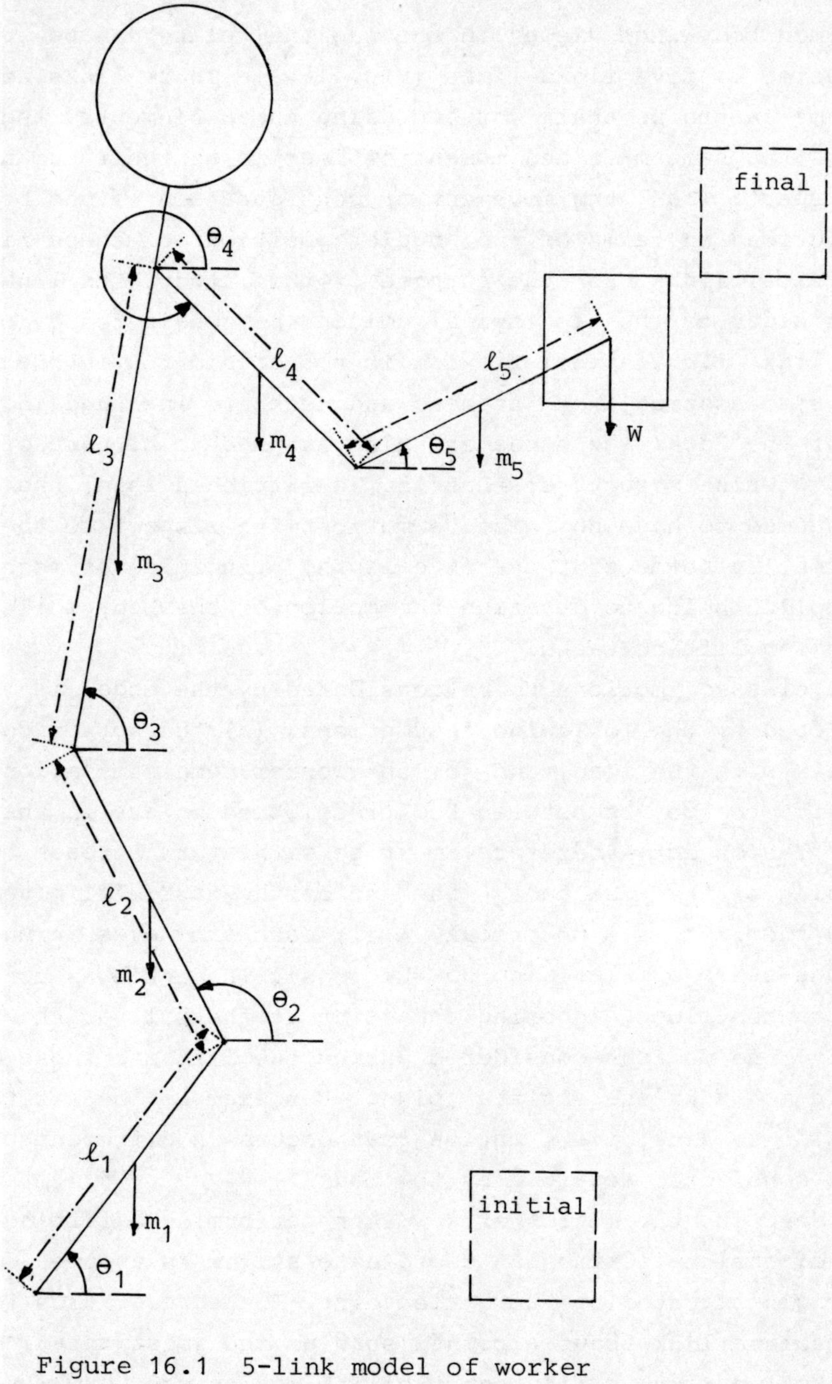


Figure 16.1 5-link model of worker

$$a_x(t) = -\dot{\Theta}^2(t)\ell \cos \Theta(t) - \ddot{\Theta}(t)\ell \sin \Theta(t)$$

$$a_y(t) = -\dot{\Theta}^2(t)\ell \sin \Theta(t) + \ddot{\Theta}(t)\ell \cos \Theta(t)$$

where $\ell$ is the distance of the object from the center of rotation, $\Theta(t)$ is the angular displacement of the link, and $\dot{\Theta}(t)$, $\ddot{\Theta}(t)$ are the corresponding angular velocity and acceleration, respectively. Therefore, linear accelerations of each joint of the link model are:

$$a_{x_1}(t) = 0$$

$$a_{y_1}(t) = 0$$

$$a_{x_{i+1}}(t) = -\ell_i[\dot{\Theta}_i^2(t) \cos \Theta_i(t) + \ddot{\Theta}_i(t) \sin \Theta_i(t)] + a_{x_i}(t)$$

$$a_{y_{i+1}}(t) = a_{y_i}(t) + \ell_i[-\dot{\Theta}_i^2(t) \sin \Theta_i(t) + \ddot{\Theta}_i(t) \sin \Theta_i(t)]$$

$$i = 1,\ldots 4$$

The linear accelerations of the center of mass of the $i^{th}$ link, can be written as:

$$a_{cx_i}(t) = a_{x_i}(t) - r_i[\dot{\Theta}_i^2(t) \cos \Theta_i(t) + \ddot{\Theta}_i(t) \sin \Theta_i(t)]$$

$$i = 1,\ldots 5 \qquad (16.1)$$

$$a_{cy_i}(t) = a_{y_i}(t) + r_i[-\dot{\Theta}_i^2(t) \sin \Theta_i(t) + \ddot{\Theta}_i(t) \sin \Theta_i(t)]$$

where $r_i$ is the distance of the center of mass from the proximal point of the $i^{th}$ link.

In order to lift an object, a worker exerts a force upon the object, and an equal and opposite force is exerted upon the worker, which is transmitted through his joints to the surface upon which he is standing. At equilibrium the sum of the forces at each joint are zero (Fig. 16.2 describes this situation for the forearm-hand link), implying:

$$F_{x_5}(t) = -m_5 a_{cx_5}(t)$$

$$F_{y_5}(t) = W + m_5 a_{cy_5}(t) + gm_5 \qquad (16.2)$$

$$F_{x_i}(t) = F_{x_{i+1}}(t) - m_i a_{cx_i}(t)$$
$$F_{y_i}(t) = F_{y_{i+1}}(t) + m_i g + m_i a_{cy_i}(t) \qquad i = 1,...,4$$

where $m_i$ is the mass of the $i^{th}$ link, and W is the weight of the object to be lifted. In addition, the torques induced at each joint are related to $\Theta_i$ by the following system.

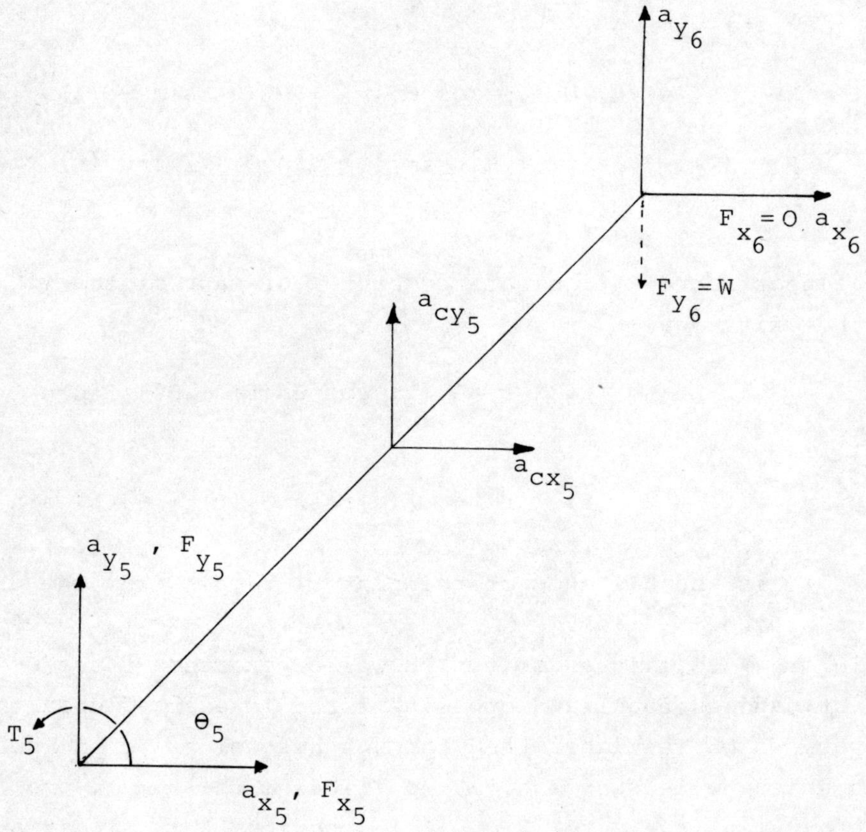


Figure 16.2 Free body diagram for elbow to wrist link system

$$T_6(t) = 0$$

$$-I_i \ddot{\Theta}_i(t) = T_{i+1}(t) - T_i(t) + m_i r_i g \cos \Theta_i(t) + \ell_i(F_{y_{i+1}}(t) \cos \Theta_i(t) - F_{x_{i+1}}(t) \sin \Theta_i(t)) + m_i r_i(a_{cx_i}(t) \sin \Theta_i(t) + a_{cy_i}(t) \cos \Theta_i(t))$$

$$i = 1,\ldots,5 \qquad (16.3)$$

where $F_{x_6}(t) = 0$, $F_{y_6}(t) = W$, g is the acceleration due to gravity, and $I_i$ is the moment of inertia for the $i^{th}$ link.

Constraints are imposed upon the lifting motion due to limitations of the human body. For example joints can only bend in one direction and by limited amounts and rates. This results in magnitude restrictions on the angles

$$l_{1i} \le \Theta_i(t) \le u_{1i}, \quad t \in [0,T], \qquad i = 1,\ldots,5 \qquad (16.4)$$

and their derivatives

$$\dot{\Theta}_i(t) \le u_{2i}, \quad \ddot{\Theta}_i(t) \le u_{3i}, \qquad i = 1,\ldots,5$$

The task itself also imposes constraints upon the motion, because the object to be lifted has an initial position and a desired final position. The worker is assumed to have an initial position such that his hands reach the object at its initial position, and this initial position of the body must be feasible according to (16.4). The worker's initial position, then, is given by:

$$\Theta_i(0) = \Theta_{io} \qquad i = 1,\ldots,5 \qquad (16.5)$$

This initial position for the worker is such that his hands grasp the object, and either his back is straight and his knees are bent, or his back is bent and his knees are straight. Since the object has a desired final condition, the worker must be able to reach it at this point. This results in a constraint on the final position of the worker:

$$\Theta_i(T) = \Theta_{iT} \qquad i = 1,\ldots,5 \qquad (16.6)$$

The final position is chosen such that the worker is standing up with his arms extended and his back bent, if needed, so as to allow his hands to reach the object.

## 16.3 FORMULATION OF THE OBJECTIVE

Nubar and Contini's Principle of Optimality for Biomechanics [3] states that a worker will minimize physical effort by moving in a trajectory consistent with the task constraints. This physical effort in performing a lifting task can be expressed as the worker's mechanical energy expenditure,a value which can be derived for this model. An approximation of it is used as the objective function. By minimizing this criterion, we invoke the Principle of Optimality as stated by Nubar and Contini.

Tichauer [4] showed that the net moments induced about the joints of the body while performing a manual task are the major indices of injury during a task. Thus, in order to minimize the possibility of an injury while performing a task, we shall determine the optimal performance of a task in the sense of Nubar and Contini since this minimum mechanical energy expenditure is directly proportional to the magnitude of the net moments induced.

Mechanical energy expenditure can be expressed as the time integral of the product of the force generated by the muscle in causing rotation and of its velocity in shortening. This fact was used by Nubar and Contini and, subsequently, Chow and Jacobson [5] to derive the criterion

$$\int_{0}^{T} \sum_{i=1}^{5} \gamma_i T_i^2(t)\,dt \qquad (16.7)$$

where $T_i(t)$ is the net torque acting at the $i^{th}$ link, and $\gamma_i$ is a weighting factor. However, since the torque at the ankle includes the effect of all net moments of the other joints, we choose to consider only this torque in the objective. As shown in Section 16.3, results obtained by use of this simplified objective agree closely with experimentally observed lifts. Thus, we represent the objective for this model as

$$F(T) = \int_0^T T_1(t)^2 \, dt \qquad (16.8)$$

where, using equations (16.1) and (16.2), $T_1$ is expressed in terms of the functions $\Theta_i(\cdot)$ as follows:

$$\begin{aligned} T_1(t) = & \sum_{i=1}^{5} [m_i r_i g \cos\Theta_i(t) + I_i \ddot{\Theta}_i(t) \\ & + \ell_i \cos\Theta_i(t) ( W + \sum_{j=i+1}^{5} m_j g )] \\ & + \sum_{i=2}^{5} m_i r_i \sum_{j=1}^{i-1} \ell_j (v_{ji}(t) - w_{ji}(t)) \\ & + \sum_{j=1}^{4} \ell_j \sum_{i=j+1}^{5} m_i r_i (v_{ij}(t) - w_{ij}(t)) \\ & + \sum_{i=1}^{4} i \sum_{j=i+1}^{5} m_j \sum_{k=1}^{j-1} \ell_k (v_{ki}(t) - w_{ki}(t)) , \end{aligned} \qquad (16.9)$$

$$v_{ji}(t) = \ddot{\Theta}_j(t) \cos ( \Theta_i(t) + \Theta_j(t) ) ,$$

$$w_{ji}(t) = \ddot{\Theta}_j(t) \sin ( \Theta_i(t) + \Theta_j(t) ) .$$

Thus we have the following control problem:

$$\text{Minimize } F(T) = \int_0^T T_1(t)^2 dt \qquad (16.10)$$

with controls $T_1(t),\ldots,T_6(t)$ and states $\Theta_1(t),\ldots,\Theta_6(t)$ subject to the differential equations (16.3), the boundary condition (16.5), (16.6) and inequality constraints (16.4).

## 16.4 NUMERICAL RESULTS

Instead of solving necessary conditions for the original minimization problem, we approximate the states $\Theta_i(\cdot)$ by a linear combination of special functions $\phi_i(\cdot)$ which are twice continuously differentiable:

$$\Theta_i(t) = \sum_{j=1}^{N} c_{ij}\phi_j(t), \; t \in [0,T] , \quad i=1,\ldots,5. \qquad (16.11)$$

By substituting (16.11) into (16.3) and (16.10), we obtain the finite dimensional minimization problem of selecting the parameter vector

$$(c_{11}, c_{12}, \dots, c_{1N}, c_{2N}, \dots, c_{5N}) \in \mathbf{R}^{5N}$$

which minimizes (16.11), subject to the constraints (16.5) - (16.7). Similarly, the constraints may be written in terms of $\phi_j(t)$ and $c_{ij}$.

The approximating functions are based on the work of Slots and Stone [ 8 ], who specify the displacement-time curve of the free joint of the body as:

$$\Theta_i(t) = \Theta_i(0) + \max(\Theta_i)/2\pi[2\pi t/T) - \sin(2\pi t/T)] \qquad (16.12)$$

where $\max(\Theta_i)$ is the maximum displacement of $\Theta_i(t)$ for $t \in [0,T]$. Although the functions (16.10) do not constitute a basis for $C^2(0,T)$, as usually assumed from Ritz approach, they reflect the specific type of motion being studied in this model. These curves also are very easy to bound in order to satisfy the constraints of the model. First, the displacement-time curves trivally satisfy the initial and final time constraints (16.5) and (16.6). Secondly, the inequality constraints

$$\begin{aligned} a_1 \le \Theta_i(t) &\le b_1 \\ \dot{\Theta}_i(t) &\le b_2 \\ \ddot{\Theta}_i(t) &\le b_3 \end{aligned} \qquad (16.13)$$

can be enforced by bounding the parameters $c_{ij}$ for each curve [ 1 ].

The discretized optimization model consists of the minimization of a nonlinear objective over a finite-dimensional space, subject to linear constraints. An algorithm based on a projected gradient modification of variable metric descent was selected. The descent was implemented by the BFGS variable metric correction. Treatment of the inequalities utilized an active set strategy and the one-dimensional search was implemented by cubic interpolation using derivatives.

The optimization model was solved on a IBM 370/165 computer using double precision wordlength.

Approximately 300K bytes of storage were required. Computation times were on the order of 5-10 minutes for a single optimization of the model.

The model and solution algorithm presented in the preceding sections were applied to solve the following manual lifting problem. A worker is required to lift a box weighing x kg and measuring s·h·1 meters. The worker has to move between two points previously selected in the work place. The time for the entire task is T seconds. The worker can use one of two methods of lift: straight knees/bent back or bent knees/straight back. The problem is to determine an optimal lift: a lift that would be performed by an individual optimizing his performance consistent with task constraints.

It is assumed that the worker is an average individual with anthropometric characteristics as given in Table 16.2. The dimensions of the box are given in Table 16.1. If handles are present on the box, these are assumed to be on the side and the worker would use them during the course of lifting. Otherwise, it is assumed that the worker will place his hands around the lower part of the box. The initial and final configurations of the body are determined by the model. Bounds on the body angles are given in Table 16.3.

| | |
|---|---|
| Width | 0.25 m |
| Length | 0.31 m |
| Depth | 0.36 m |
| Weight | 6.1 kg |

Table 16.1 Box Dimensions

| Body Segment | Length | Mass (Bodymass=M) | Moment of Inertia | Center of Mass |
|---|---|---|---|---|
| Ankle to knee | $L_1 = 0.41$m | $.046M = M_1$ | $M_1L_1^2(.5565^2 - .438^2)$ | $.433L_1$ |
| Knee to hip | $L_2 = 0.43$m | $.105M = M_2$ | $M_2L_2^2(.542^2 - .43^2)$ | $.43L_2$ |
| Hip to shoulder | $L_3 = 0.46$m | $.554M = M_3$ | $M_3L_3^2(.83^2 - .66^2)$ | $.66L_3$ |
| Shoulder to elbow | $L_4 = 0.25$m | $.031M = M_4$ | $M_4L_4^2(.54^2 - .43^2)$ | $.43L_4$ |
| Elbow to wrist | $L_5 = 0.23$m | $.025M = M_5$ | $M_5L_5^2(M_5L_5^2 - .443^2)$ | $.438L_5$ |

Note: The hipspan is 0.35m and the total body weight is 75 kg

Table 16.2

| Joint | Bounds (in degrees) |
|---|---|
| ankle | $40 \le \theta_1(t) \le 90$ |
| knee | $90 \le \theta_2(t) \le 225$ |
| hip | $-10 \le \theta_3(t) \le 110$ |
| shoulder | $-95 \le \theta_4(t) \le 90$ |
| elbow | $-90 \le \theta_5(t) \le 90$ |

Table 16.3 Bounds on Angles

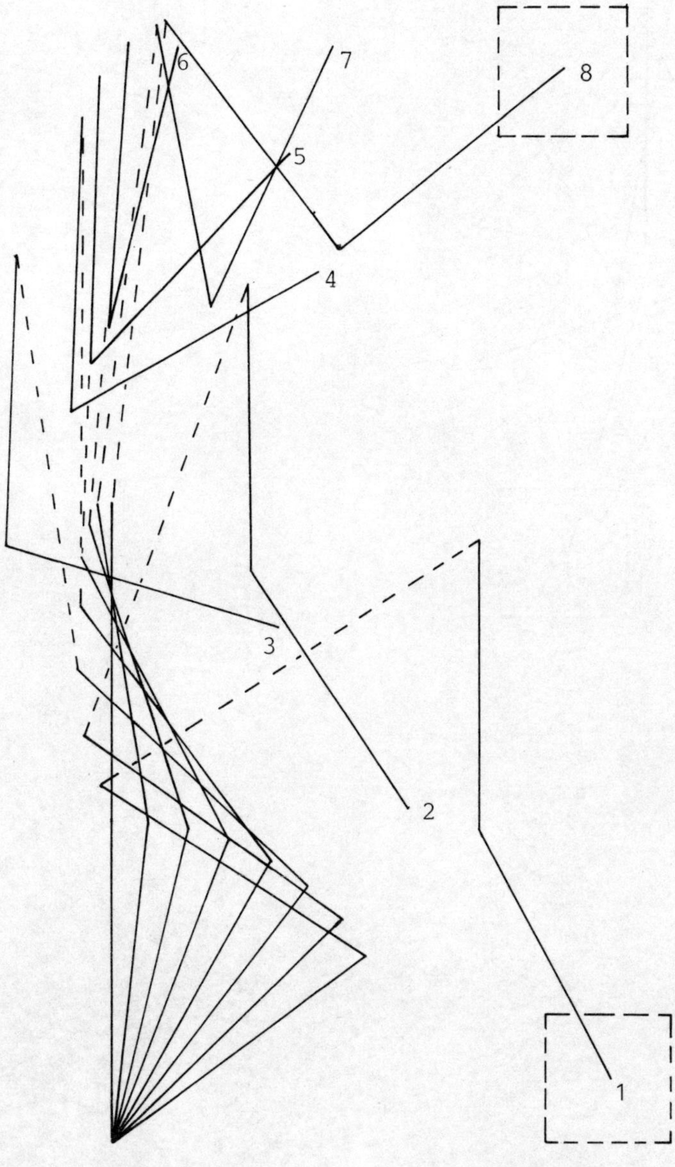


Figure 16.3 Foot to shoulder lift

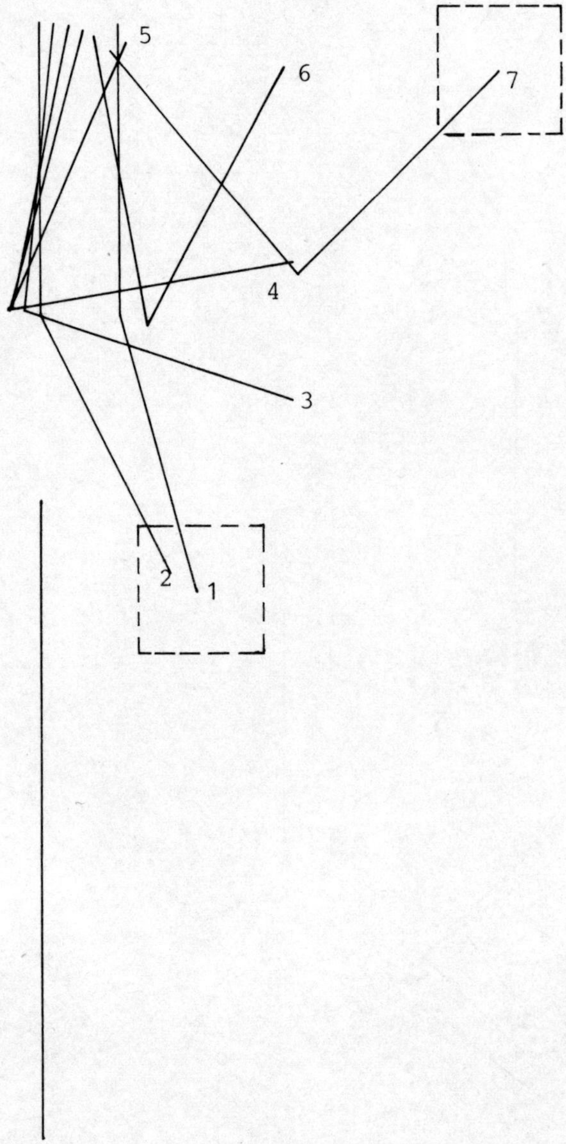


Figure 16.4 Knuckle to shoulder lift

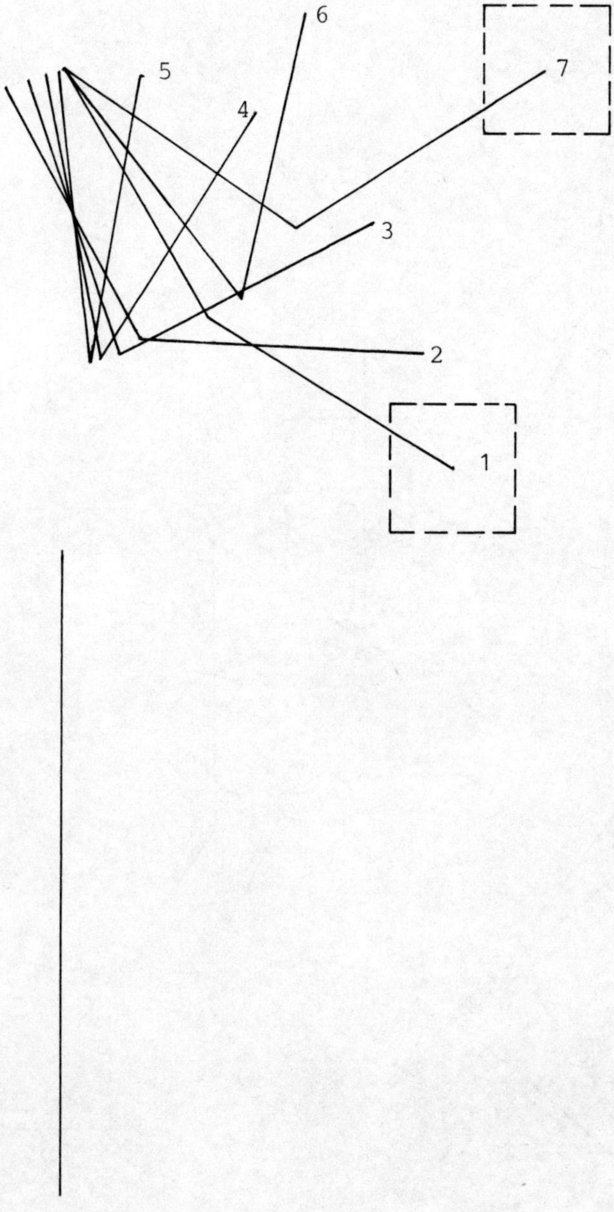


Figure 16.5 Waist to shoulder lift

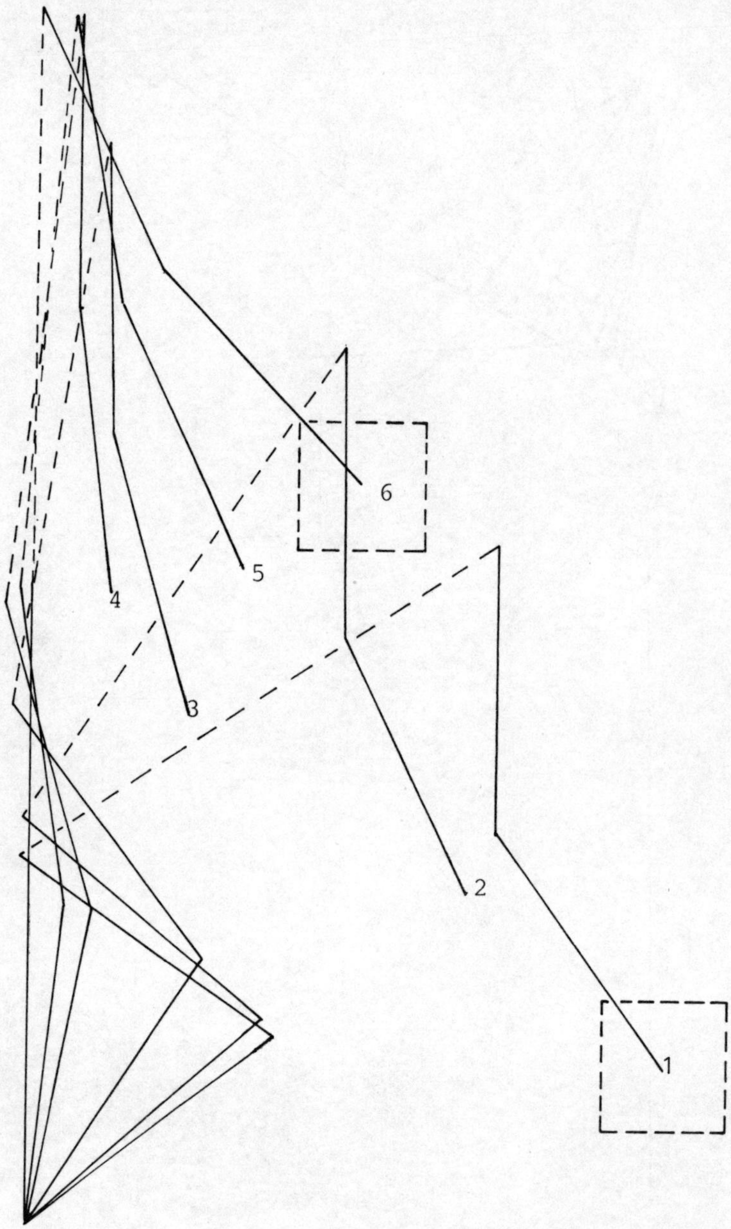


Figure 16.6 Foot to waist lift

The model was solved for three different lifts: (1) foot to shoulder, (2) waist to shoulder, and (3) foot to waist. For all three lifting tasks, it was assumed that handles were available on the side of the box for being grasped by the worker.

Foot to Shoulder Lift: For this method, the worker is assumed to lift the box from his feet to his shoulders. The first experiment assumes the worker has 3 seconds to lift the box using the bent knees/straight-back method. The optimal cost for the worker is 144.82 $(N\text{-}m)^2$; a diagram of the motion is shown in Fig. 16.3 . Using the straight knees/bent back method, the optimal cost is 126.2 $(N\text{-}m)^2$; see Fig. 16.4 . However, the straight knees/bent back method of lift requires the box to be 8 inches off the floor initially because the worker cannot reach the floor. Finally, the worker is allowed only 2 seconds to lift the box. Using the bent knees/straight back method, the optimal cost is higher, 202.67 $(N\text{-}m)^2$.

Waist to Shoulder Lift: This method assumes that the worker lifts from waist to shoulder height. Because the worker remains standing, only the straight knees/bent back lift was studied. For the first run with a time of 2 seconds, the task costs the worker 51.29 $(N\text{-}m)^2$; the optimal motion is shown in Fig. 16.5 . The second run assumes only 1 second for the lift, and cost is 41.33 $(N\text{-}m)^2$.

Foot to Waist Lift: For this experiment the worker lifts the box from his feet to his waist. Using the straight knees/back bent method of lift, the box is 8 inches off the floor initially and the model is solved for final times of 1 and 2 seconds; the optimal costs to the worker are 16.44 $(N\text{-}m)^2$ and 19.238 $(N\text{-}m)^2$, respectively. When the method of lift is bent knees/straight back, the optimal cost is 75.57 $(N\text{-}m)^2$ to lift the box in 2 seconds (Fig. 16.6 ) and 68.504 $(N\text{-}m)^2$ to lift the box in 1 second.

The paths of motions predicted by the model when using either method of lift (bent knees/straight back and straight knees/bent back) are consistent with the results of laboratory experiments. A box or an object is usually lifted by quickly bringing the weight of the object as close to the hips as possible, and then thrusting the object outward and setting it down where desired. This is the action predicted by the model.

The time allowed for the lift is a significant factor in determining the cost of the task. For the foot to shoulder lift, using bent knees/straight back, an increase in time from 2 seconds to 3 seconds results in a decreased cost for the task. With the shorter time interval, the worker has to move faster throughout the entire lift, and this causes increased optimal cost. The reverse is true for the waist to shoulder lift and the foot to waist lift, both of which uses the bent knees/ straight back method: a decrease in time from 2 seconds to 1 second results in a decreased cost for the task. One possible reason for that decrease is that the actual distance, through which the box is lifted, is small enough that no savings results from quickly moving the box close to the body and holding it there as long as possible before thrusting it out since such an action results in more acceleration at critical parts of the lift. The choice of time for a task is critical in determining the cost of the task since it influences the speed and acceleration at which the task is accomplished.

For the foot to waist lift, the bent knees/straight back method costs more than the straight knees/bent back method. This reflects the fact that using the straight knees/bent back method does not induce any torques about the knee in contrast to the bent knees/straight back method. Also, a comparison of the results for the two methods of lifting is inconclusive for both the foot to shoulder task and the foot to waist task because, for the straight knees/bent back method, the box was not as low initially as in the case of the straight back/bent knees method.

## 16.5 COMMENTS AND REFERENCES

Section 16.1: The experimental approach for analyzing lifting problems has been considered in [9] - [15]. Some models, such as that considered by Martin and Chaffin [15] also take special flexion into account. The approach presented in this chapter has also been carried out for lifting problems in which obstacles limit the motion [1, 2].

Section 16.2: Chow and Jacobson [5] also consider a simplified optimal control approach for determining optimal human motion such as walking. An optimal control model for the kip-up movement was analyzed in [16].

Section 16.3: The problem reformulation given in this section can also be viewed as one of the calculus of variations [6] by substituting the expression for $T_1^2$ in (16.3) into (16.9), obtaining an unconstrained functional depending on $\Theta$, $\dot{\Theta}$ and $\ddot{\Theta}$. A necessary condition for optimality is that the Euler-Lagrange equation be satisfied subject to the endpoint condition. Such problems can be solved also by shooting methods, for example. The additional requirement of inequality constraints on $\Theta(t)$ requires that the Lagrange multipliers be discontinuous at points where the constraints become active [7].

Section 16.4: Inequality constraints were accommodated in the projection-restoration method using an active set strategy due to Murtagh and Sargent [17].

# References

CHAPTER 2

1 M. L. Lenard, Practical convergence conditions for unconstrained optimization, Math. Programming 4 (1973), 309-323.

2 D. F. Shanno, On the convergence of a new conjugate gradient algorithm, SIAM J. Numer. Anal. 15 (1978), 1247-1257.

3 G. Zoutendijk, cf. E. Polak, Computational methods in optimization (New York: Academic Press, 1971), p.48.

4 H. B. Curry, The method of steepest descent for nonlinear minimization problems. Quart. Appl. Math. 2 (1944), 258-263.

5 L. Armijo, Minimization of functions having Lipschitz-continuous first partial derivatives. Pacific J. Math. 16 (1966), 1-3.

6 A. A. Goldstein, Constructive real analysis. (New York: Harper, 1967).

7 M. J. D. Powell, Some global convergence properties of a variable metric algorithm for minimization without exact line searches. In: Nonlinear Programming, SIAM-AMS Proceedings 9, Providence, Amer. Math. Soc. 1976.

8 J. W. Daniel, The approximate minimization of functionals. (Englewood Cliffs: Prentice-Hall, 1971).

9 E. Polak Computational methods in optimization, a unified approach.(New York: Academic Press 1971).

10 W. I. Zangwill, Nonlinear programming: a unified approach. (Englewood Cliffs: Prentice-Hall, 1969).

CHAPTER 3

1 W. C. Davidon, Variable metric method for minimization. Argonne National Laboratories, Rep. ANL-5990 (1959).

2 R. Fletcher and M. J. D. Powell, A rapidly convergent descent method for minimization. Comput. J. 6 (1963), 163-168.

3 J. E. Dennis and J. J. Moré, Quasi-Newton methods, motivation and theory. SIAM Rev. 19 (1977), 46-89.

4 C. G. Broyden, Quasi-Newton methods and their application to function minimisation. Math. Comp. 21 (1967), 368-381.

5 R. Fletcher, A new approach to variable metric algorithms. Comput. J. 13 (1970), 317-322.

6 H. Y. Huang, Unified approach to quadratically convergent algorithms for function minimization. J. Optim. Theory Appl. 5 (1970), 405-423.

7 K. Ritter, Global and superlinear convergence of a class of variable metric methods. University of Wisconsin (Madison), MRC-Report 1945 (1979).

8 S. S. Oren and D. G. Luenberger, Self-scaling variable metric (SSVM) algorithms, part I: criteria and sufficient conditions for scaling a class of algorithms. Management Sci. 20 (1974), 845-862.

9 S. S. Oren, Self-scaling variable metric (SSVM) algorithms part II : implementation and experiments. Management Sci. 20 (1974), 863-874.

10 C. G. Broyden, A new double rank minimization algorithm. Notices Amer. Math. Soc. 16 (1969), 670.

11 D. Goldfarb, A family of variable-metric methods derived by variational means. Math. Comp. 24 (1970), 23-26.

12 D. F. Shanno, Conditioning of quasi-Newton methods for function minimization. Math. Comp. 24 (1970), 647-656.

CHAPTER 4

1 J. E. Dennis and J. J. Moré, Quasi-Newton methods, motivation and theory. SIAM Rev. 19 (1977), 46-89.

2 W. Warth and J. Werner, Effiziente Schrittweitenfunktionen bei unrestringierten Optimierungsaufgaben. Computing 19 (1977), 59-72.

3 J. Werner, Über die globale Konvergenz von Variable-Metrik-Verfahren mit nicht-exakter Schrittweitenbestimmung. Numer. Math. 31 (1978), 321-334.

4 S. S. Oren and E. Spedicato, Optimal conditioning of self-scaling variable metric algorithms. Math. Programming 10 (1976), 70-90.

5 M. J. D. Powell, Some global convergence properties of a variable metric algorithm for minimization without exact line searches. In : Nonlinear Programming, SIAM-AMS Proceedings 9, Providence, Amer. Math. Soc. 1976.

6 L. B. Horwitz abd P. E. Sarachik, Davidon's method in Hilbert space, SIAM J. Appl. Math. 16 (1968), 676-695.

7 H. Tokumaru, N. Adachi and K. Goto, Davidon's method for minimization problems in Hilbert space with an application to control problems. SIAM J. Control Optim. 8 (1970), 163-178.

8 P. R. Turner and E. Huntley, Direct-prediction quasi-Newton methods in Hilbert space with applications to control problems. J. Optim. Theory Appl. 21

(1977), 199-211.

9 P. R. Turner and E. Huntley, Variable metric methods in Hilbert space with applications to control problems. J. Optim. Theory Appl. 19 (1976), 381-400.

10 H. Y. Huang, Unified approach to quadratically convergent algorithms for function minimization. J. Optim. Theory Appl. 5 (1970), 405-423.

11 E. R. Edge and W. F. Powers, Function-space quasi-Newton algorithms for optimal control problems with bounded controls and singular arcs. J. Optim. Theory Appl. 20 (1976), 455-479.

CHAPTER 5

1 J. E. Dennis and J. J. Moré, A characterization of superlinear convergence and its application to quasi-Newton methods. Math. Comp. 28 (1974), 549-560.

2 G. G. Broyden, J. E. Dennis and J. J. Moré, On the local and superlinear convergence of quasi-Newton methods. J. Inst. Math. Appl. 12 (1973), 223-246.

3 J. E. Dennis and J. J. Moré, Quasi-Newton methods, motivation and theory. SIAM Rev. 19 (1977), 46-89.

4 M. J. D. Powell, Some global convergence properties of a variable metric algorithm for minimization without exact line searches, in Nonlinear programming, SIAM-AMS Proceedings 9, Providence, Amer. Math. Soc. 1976.

5 M. L. Lenard, Practical convergence conditions for the Davidon-Fletcher-Powell method. Math. Programming 9 (1975), 69-86.

6 J. Stoer, On the convergence rate of imperfect minimization algorithms in Broyden's β-class. Math. Programming 9 (1975), 313-335.

7 J. Werner, Über die globale Konvergenz von Variable-Metrik-Verfahren mit nicht-exakter Schrittweitenbestimmung. Numer. Math. 31 (1978), 321-334.

8 K. Ritter Local and superlinear convergence of a class of variable metric methods. Computing 23 (1979), 287-297.

9 K. Ritter On the rate of superlinear convergence of a class of variable metric methods. University of Wisconsin (Madison), MRC-Report 1949 (1979).

10 K. Ritter, Global and superlinear convergence of a class of scaled variable metric methods. University of Madison (Madison), MRC-Report 1967 (1979).

CHAPTER 6


1 M. R. Hestenes and E. Stiefel, Methods of conjugate gradients for solving linear systems. J. Res. Nat. Bur. Standards 49 (1952), 409-436.

2 R. Fletcher and C. M. Reeves, Function minimization by conjugate gradients. Computer J. 7 (1964), 149-154.

3 M. L. Lenard, Practical convergence conditions for unconstrained optimization. Math. Programming 4 (1973), 309-323.

4 D. F. Shanno, Conjugate gradient methods with inexact searches. Math. Oper. Res. 3 (1978), 244-256.

5 D. F. Shanno, On the convergence of a new conjugate gradient algorithm. SIAM J. Numer. Anal. 15 (1978), 1247-1257.

6 L. Nazareth, A family of variable metric updates. Math. Programming 12 (1977), 157-172.

7 G. P. McCormick and K. Ritter, Alternative proofs of the convergence properties of the conjugate-gradient method. J. Optim. Theory Appl. 13 (1974), 497-518.

8 W. Warth and J. Werner, Effiziente Schrittweitenfunktionen bei unrestringierten Optimierungsaufgaben. Computing 19 (1977), 59-72.

9 J. W. Daniel, The conjugate gradient method for linear and nonlinear operator equations. SIAM J. Numer. Anal. 4 (1967), 10-26.

10 J. W. Daniel, The approximate minimization of functionals. (Englewood Cliffs: Prentice-Hall, 1971).

11 Z. Fortuna, Superlinear convergence of conjugate gradient method. Control Cybernet. 3 (1974), 63-68.

12 Z. Fortuna, Some convergence properties of the conjugate gradient method in Hilbert space. SIAM J. Numer. Anal. 16 (1979), 380-384.

## CHAPTER 7


1 M. Frank and P. Wolfe, An algorithm for quadratic programming. Naval Res. Logist. Quart. 3 (1956), 95-110.

2 V. F. Demyanov and A. M. Rubinov, On the problem of minimization of a smooth functional with convex constraints. Soviet Math. Dokl. 6 (1965), 9-11.

3 M. Valadier, Extension d'un algorithme de Frank et Wolfe. Rev. Franc. Réch. Opér. 36 (1965), 251-253.

4 E. G. Gilbert, An iterative procedure for computing the minimum of a quadratic form on a convex set. SIAM J. Control Optim. 4 (1966), 61-80.

5 R. O. Barr, An efficient computational procedure for a generalized quadratic programming problem. SIAM J. Control Optim. 7 (1969), 415-429.

A. Auslender and F. Brodeau, Convergence d'un algorithme de Frank et Wolfe appliqué á un problème de contrôle. RAIRO Réch. Opér. 7 (1968), 3-12.

7 J. C. Dunn and S. Harshbarger, Conditional gradient algorithms with open loop step size rules. J. Math. Anal. Appl. 62 (1978), 432-444.

8 M. D. Canon and C. D. Cullum, A tight upper bound on the rate of convergence of the Frank-Wolfe algorithm. SIAM J. Control Optim. 6 (1968), 509-516.

9 J. C. Dunn, Rates of convergence for conditional gradient algorithms near singular and nonsingular ex-

tremals. SIAM. J. Control Optim. 17 (1979), 187-211.

10 E. Sachs, Differenzierbarkeit in der Optimierungstheorie und Anwendungen auf Kontrollprobleme. (Technische Hochschule Darmstadt, Dissertation, 1975).

11 D. Q. Mayne and E. Polak, First-order strong variation algorithms for optimal control. J. Optim. Theory Appl. 16 (1975), 277-301.

12 D. Q. Mayne and E. Polak, First-order strong variation algorithms for optimal control problems with terminal inequality constraints. J. Optim. Theory Appl. 16 (1975), 303-325.

CHAPTER 8

1 H. Uzawa, Iterative methods for concave programming. In: Studies in linear and nonlinear programming, eds Arrow, Hurwicz, Uzawa. (Stanford: University press, 1958).

2 J. B. Rosen, The gradient projection method for nonlinear programming I : linear constraints , II: nonlinear constraints. SIAM J. Appl. Math. 8 (1960),180-217; 9 (1961), 514-532.

3 V. F. Demyanov and A. M. Rubinov, Approximate methods in optimization problems. (New York: Elsevier, 1970).

4 J. W. Daniel, The approximate minimization of functionals. (Englewood Cliffs: Prentice-Hall, 1971).

5 E. S. Levitin and B. T. Polyak, Constrained minimization methods. USSR Comput. Math. and Math. Phys. 6 (1966), 1-50.

CHAPTER 9


1 D. Wulbert, Uniqueness and differential characterization of approximation from manifolds of functions. Amer. J. Math. 93 (1971), 350-366.

2 W. Cheney, Introduction to approximation theory. (New York: McGraw-Hill, 1966).

3 M. R. Osborne and G. A. Watson, An algorithm for minimax approximation in the nonlinear case. Comput. J. 12 (1969), 64-69.

4 M. R. Osborne and G. A. Watson, Nonlinear approximation problems in vector norms. In Numerical analysis, ed. G. A. Watson. (Berlin: Springer,1978), 117-132.

5 K. Madsen, Minimax solution of non-linear equations without calculating derivatives. Math. Programming Stud. 3 (1975), 110-126.

6 J. Hald and H. Schjaer-Jacobsen, Linearly constrained minimax optimization without calculating derivatives. Operations Research Verfahren 31 (1979), 289-301.

7 L. Cromme, Eine Klasse von Verfahren zur Ermittlung bester nichtlinearer Tschebyscheff-Approximationen. Numer. Math. 25 (1975), 447-459.

8 L. Cromme, Numerische Methoden zur Behandlung einiger Problemklassen der nichtlinearen Tschebyscheff-Approximation mit Nebenbedingungen. Numer. Math. 28 (1977), 101-117.

9 R. Reemtsen, On the convergence of a class of nonlinear approximation methods. Technische Hochschule Darmstadt, Preprint-Nr. 536 (1980).

10 R. Reemtsen, A computer program for the numerical solution of free boundary problems. University of Delaware, Applied Mathematics Institute, Technical report no. 56 A (1979).

11 P. Jochum, Optimale Kontrolle von Stefan-Problemen mit Methoden der nichtlinearen Approximationstheorie. (Universität München, Dissertation, 1978).

CHAPTER 10


1 H. S. Carslaw and J. C. Jaeger, Conduction of heat in solids. 2nd ed. (Oxford: University Press, 1959).

2 E. Sachs, Optimal control for a class of integral equations, in Optimization and Operations Research Proceedings. (Berlin: Springer, 1978).

3 E. Sachs, A parabolic control problem with a boundary condition of the Stefan-Boltzmann type. Z. Angew. Math. Mech. 58 (1978), 443-449.

4 S. G. Michlin, Vorlesungen über lineare Integralgleichungen. (Berlin: Verlag der Wissenschaften, 1962).

5 S. G. Michlin, The numerical performance of variational methods. (Groningen: Noordhoff, 1971).

6 L. Collatz and W. Krabs, Approximationstheorie. (Stuttgart: Teubner, 1973).

7 D. Braess, Kritische Punkte bei der nichtlinearen Tschebyscheff-Approximation. Math. Z. 132 (1973), 327-341.

8 D. Wulbert, Uniqueness and differential characterization of approximation from manifolds of functions, Amer. J. Math. 93 (1971), 350-366.

9 A. Karafiat, The problem of the number of switches in parabolic equations with control. Ann. Polon. Math. 34 (1977), 289-316.

10 A. G. Butkoskiy, Distributed control systems. (New York: Elsevier, 1969).

11 L. v. Wolfersdorf, Optimale Steuerung einer Klasse nichtlinearer Aufheizungsprozesse. Z. Angew. Math. Mech. 55 (1975), 353-362.

12 A. Friedman, Optimal control for parabolic equations. J. Math. Anal. Appl. 18 (1967), 479-491.

13 W. Cheney, Introduction to approximation theory. (New York: McGraw-Hill, 1966).

14 S. Dolecki and D. L. Russell, A general theory of observation and control. SIAM J. Control Optim. 15 (1977), 185-220.

15 P. Jochum, Optimale Kontrolle von Stefan-Problemen mit Methoden der nichtlinearen Approximationstheorie. (Universität München, Dissertation, 1978).

16 J.-P. Yvon, Etude de quelques problèmes de contrôle pour les systèmes distribués. (Université Paris, Thèse, 1973).

CHAPTER 11


1 K. Glashoff and E. Sachs, On theoretical and numerical aspects of the bang-bang-principle. Numer. Math. 29 (1977), 93-113.

2 B. N. Pshenichnyi, Necessary conditions for an extremum. (New York: Dekker, 1971).

3 K. Glashoff and N. Weck, Boundary control of parabolic differential equations in arbitrary dimensions: supremum-norm problems. SIAM J. Control Optim. 14 (1976), 662-682.

4 A. Karafiat, The problem of the number of switches in parabolic equations with control. Ann. Polon. Math. 34 (1977), 289-316.

5 F. R. Gantmacher and M. G. Krein, Oszillationsmatrizen, Oszillationskerne und kleine Schwingungen mechanischer Systeme. (Berlin: Akademie-Verlag 1960).

6 W. Cheney, Introduction to approximation theory. (New York: McGraw-Hill, 1966).

7 E. Sachs, Differenzierbarkeit in der Optimierungstheorie und Anwendung auf Kontrollprobleme. (Technische Hochschule Darmstadt, Dissertation, 1975).

8 M. Denn, Optimization by variational methods. (New York: McGraw-Hill, 1969).

9 Y. Sakawa, Solution of an optimal control problem in a distributed parameter system. IEEE Trans.

Autom. Control 9 (1964), 420-426.

10 K. Glashoff and S.-A. Gustafson, Numerical treatment of a parabolic boundary-value control problem. J. Optim. Theory Appl. 19 (1979), 645-663.

11 Y. V. Yegorov, Some problems in the theory of optimal control. U.S.S.R. Computational Math. and Math. Phys. 3 (1963), 1209-1232.

12 K. Glashoff, Über Kontrollprobleme bei parabolischen Anfangsrandwertaufgaben. (Technische Hochschule Darmstadt, Habilitationsschrift, 1975).

13 A. G. Butkovskiy, Distributed control systems. (New York: Elsevier, 1969).

14 A. Bossavit, A linear control problem for a system governed by a partial differential equation. In: Computing methods in optimization problems, eds. L. A. Zadeh et al. (New York: Academic Press, 1969), 47-54.

15 H. O. Fattorini, Boundary control systems. SIAM J. Control Optim. 6 (1968), 349-385.

16 D. L. Russell, Controllability and stabilizability for linear partial differential equations: recent progress and open questions. SIAM Rev. 20 (1978), 639-735.

17 K. Schittkowski, Numerical solution of a time-optimal parabolic boundary-value control problem. J. Optim. Theory Appl. 27 (1979), 271-290.

18 G. Knowles, Some problems in the control of distributed systems, and their numerical solution. SIAM J. Control Optim. 17 (1979), 5-22.

19 R. M. Goldwyn, K. P. Sriram and M. Graham, Time optimal control of a linear diffusion process. SIAM J. Control Optim. 5 (1967), 295-308.

20 T. Seidman and W. Chewing, A convergent scheme for boundary control of the heat equation. SIAM J. Control Optim. 15 (1977), 64-72.

CHAPTER 12


1 H. Triebel, Höhere Analysis. (Berlin: Verlag der Wissenschaften, 1972).

2 W. Eichenauer and W. Krabs, On the numerical solution of certain control-approximation problems: I. Application to the vibrating string. Technische Hochschule Darmstadt, Preprint Nr. 358 (1977).

3 W. Krabs, Über die einseitige Randsteuerung einer schwingenden Saite in einen Zustand minimaler Energie. Computing 17 (1977), 351-359.

4 J. Warga, Optimal control of differential and functional equations. (New York: Academic Press, 1972).

5 W. Eichenauer and W. Krabs, On the numerical solution of certain control-approximation problems III. Technische Hochschule Darmstadt, Preprint Nr. 470 (1979).

6 D. L. Russell, Controllability and stabilizability for linear partial differential equations: recent progress and open questions. SIAM Rev. 20 (1978), 639-735.

7 M. Hilpert, Zur Diskretisierung eines Kontroll-Approximationsproblems. Universität Berlin, Preprint Nr. 83 (1979).

8 R. M. Goldwyn, K. P. Sriram and M. Graham, Time optimal control of a linear hyperbolic system. Internat. J. Control 12 (1970), 645-656.

9 G. Knowles, Some problems in the control of distributed systems, and their numerical solution. SIAM J. Control Optim. 17 (1979), 5-22.

10 W. Eichenauer, Zur optimalen Randkontrolle eines schwingenden Balkens, Z. Angew. Math. Mech. 59 (1979), T 94 - T 96.

11 W. Eichenauer and W. Krabs, On a constructive method for solving certain control-approximation problems. In Approximation in Theorie und Praxis, ed. Meinardus (Mannheim : BI-Verlag, 1979).

CHAPTER 13

1 W. A. Gruver and N. H. Engersbach, Nonlinear programming by projection-restoration applied to optimal geostationary satellite positioning. AIAA J. 12 (1974), 1715-1720.

2 W. A. Gruver and N. H. Engersbach, Optimal impulsive trajectory rendezvous by mathematical programming. J. Comp. Meth. Appl. Mech. Engrg. 11 (1977), 165-174.

3 W. H. Goodyear, Completely general closed form solution for coordinates and partial derivatives of the two-body problem. Astronom. J. 70 (1965), 189-192, ibid. errata p. 446.

4 W. H. Goodyear, A general method for the computation of Cartesian coordinates and partial derivatives of the two-body problem, NASA report CR-522 (1966).

5 W. A. Gruver and N. H. Engersbach, A variable metric method for constrained minimization based on an augmented Lagrangian. Internat. J. Numer. Methods Engrg. 10 (1976), 1047-1056.

6 D. F. Lawden, Optimal trajectories for space navigation. (London: Butterworth, 1963).

7 P. Lion and M. Handelsman, Primer vector on fixed-time impulsive trajectories. AIAA J. 6 (1968), 127-132.

8 D. Jazweski and J. R. Doll, An efficient method for calculating optimal free-space N-impulse trajectories. AIAA J. 6 (1968), 2160-2165.

9 D. W. Gobetz and J. R. Doll, A survey of impulsive trajectories. AIAA J. 7 (1969), 127-132.

10 P. R. Escobal, Methods of Orbit Determination. (New York: Wiley, 1965).

## CHAPTER 14


1 C. F. Klein and W. A. Gruver, Optimal control of Markovian queueing systems. Optimal Control Applic. and Meth. 2 (February 1981).

2 C. F. Klein and W. A. Gruver, Dynamic optimization in Markovian queueing systems. In Differential Games and Control Theory, eds. P. T. Liu and E. Roxin (New York: Marcel Dekker, 1979), pp. 95-118.

3 D. Gross and C. M. Harris, Fundamentals of queueing theory. (New York: Wiley, 1974).

4 T. B. Grabill, D. Gross and M. J. Magazine, A classified bibliography of research on optimal design and control of queues. Oper. Res. 25 (1977), 219-232.

5 B. L. Miller, Finite state continuous-time Merkov decision processes with applications to a class of optimization problems in queueing theory, (Stanford University, Ph. D.Thesis, 1967).

6 H. Emmons, The optimal admission policy to a multiserver queue with finite horizon. J. Appl. Probab. 9 (1972), 103-116.

7 A. Martin-Lof, Optimal Control of a continuous-time Markov chain with periodic transition probabilities. Oper. Res. 15 (1967), 872-881.

8 G. P. Klimov, Extremal problems in queueing theory, in Cybernetics in the Service of Communism, Vol.2, Reliability and Theory of Queues, eds. A. I. Berg, N. G. Gruevich and V. B. Gnedenko, (Moscow: Energiza, 1964).

9 F. T. Man, Optimal control of time-varying queueing systems. Management Sci. 19 (1973), 1249-1256.

10 A. Wouk, An extension of the Carathéodory-Kalman variational method. J. Optim. Theory Appl. 3 (1969), 2-33.

11 C. E. Agnew, Dynamic modeling and control of congestion-prone systems. Oper. Res. 24 (1976), 400-419.

12 R. Feindor, Optimale Steuerung allgemeiner Warteprozesse, (Universität Karlsruhe, Doctoral Thesis, 1975).

13 Z. F. Lansdowne, The theory and applications of generalized linear control processes, (Stanford University, Master's Thesis, 1970).

14 G. Reuter and W. Lederman, On the differential equations for the transition probabilities of Markov processes with enumerably many states. Proc. Cambridge Phil. Soc. 49 (1953), 247-262.

CHAPTER 15


1 W. A. Gruver and S. L. Narasimhan, Dynamic optimization model in integrated production systems: computational aspects. IEEE Trans. Systems Man Cybernet. 10 (1979), 689-695.

2 S. L. Narasimhan and W. A. Gruver, Optimal control in integrated R&D production and inventory planning systems. AIIE Trans. 11 (1979), 198-206.

3 C. Holt, F. Modigliani , J. Muth and H. Simon, Planning production, investment and workforce. (Englewood Cliffs: Prentice-Hall,1960).

4 M. D. Intriligator, Mathematical optimization and economic theory. (Englewood Cliffs: Prentice-Hall, 1971).

5 C. I. Ferguson, Microeconomic theory. (Homewood: Irwin, 1969).

6 E. Mansfield et al., Research and innovation in the modern corporation. (New York: Norton, 1971).

7 P. T. Lele and J. W. O'Leary, Applications of production functions in management decisions. AIIE Trans. 4 (1972), 36-42.

8 W. Leontief, Structure of the american economy 1919-1939. (New York: Oxford Univ. Press, 1951).

9 R. M. Solow, Growth theory. (New York: Oxford Univ. Press, 1970).

10 V. Smith, Theory of investment and production. (Cambridge: Harvard Univ. Press, 1961).

11 Z. Griliches, Research expenditure, education and the aggregate agricultural production functions. American Econ. Rev. 54 (1964), 961-974.

12 K. J. Arrow and M. Kurz, Public investment, the rate of return and optimal fiscal policy. (Baltimore: Johns Hopkins Univ. Press, 1971).

13 S. Axsäter, Coordinating control of production-inventory systems. Int. J. Prod. Res. 14 (1976), 669-688.

14 E. S. Lee and M. A. Shaikh, Optimal production planning by a gradient technique, I. First variations. Management Sci. 16 (1969), 109-117.

15 Z. Lieber, An extension to Modigliani and Hohn's planning horizon results. Management Sci. 20 (1973)m 319-330.

16 D. Pekelman, Production smoothing with fluctuating price. Management Sci. 21 (1975), 576-590.

17 P. R. Kleindorfer , C. H. Kriebel, G. L. Thompson and G. B. Kleindorfer, Discrete optimal control of production plans. Management Sci. 22 (1975), 261-273.

18 I. Adiri and A. Ben-Israel, An extension and solution of Arrow-Karlin type production models by the Pontryagin maximum principle. Cahiers Centre Etudes Rech. Opér. 8 (1966), 147-158.

19 A. Bensoussan, E. G. Hurst and B. Näslund, Management applications of modern control theory. (Amsterdam: North-Holland, 1974).

20 S. L. Narasimhan and W. A. Gruver, Comparison of optimal control and linear programming techniques in an integrated planning environment. Proc. of the Northeast AIDS Conf. (Washington: June 1978).

CHAPTER 16

1 M. Muth, M. A. Ayoub and W. A. Gruver, A nonlinear programming model for the evaluation of manual lifting tasks, in C. G. Drury (ed.), Safety in Manual Materials Handling. (Washington: U.S. Government Printing Office (NIOSH), 1978).

2 W. A. Gruver, M. A. Ayoub and M. Muth, A model for optimal evaluation of manual lifting tasks. J. Safety Research 11 (1979), 61-71.

3 Y. Nubar and R. Contini, A minimal principle in biomechanics. Bull. Mathem. Biophysics 23 (1961), 377-391.

4 E. A. Tichauer, A pilot study of the biomechanics of lifting in simulated industrial work situations. J. Safety Research 3 (1971), 98-115.

5 C. K. Chow and D. H. Jacobson, Studies on human locomotion via optimal programming. Math. Biosci. 10 (1971), 239-306.

6 I. M. Gelfand and S. V. Fomin, Calculus of variations. (Englewood Cliffs: Prentice-Hall, 1963).

7 D. R. Smith, Variational methods in optimization. (Englewood Cliffs: Prentice-Hall, 1974).

8 L. Slots and G. Stone, Biomechanical power generated by forearm flexion. Human Factors 5 (1969), 443-452.

9 W. S. Frederick, Human energy in manual lifting. Modern Materials Handling 14 (1956).

10 U. Aberg, K. Elgstrand, P. Magnus and A. Lindham, Analysis of components and prediction of energy expenditure in manual tasks. Internat. J. Prod. 6 (1968), 189-196.

11 S. H. Snook and C. H. Irvine, Maximum acceptable weight of lift. Amer. Industr. Hygiene Assoc. J. 28 (1967), 322-329.

12 M. A. Ayoub, A biomechanical model for the upper extremity using optimization techniques. (Texas Tech University, Ph. D. Thesis, 1971).

13 C. G. Drury and R. E. Pfeil, A task-based model of manual lifting performance. Internat. J. Production Res. 13 (1975), 137-148.

14 A. Roozbazar, A work physiology study of lifting. (North Carolina State University, Ph. D. Thesis, 1974).

15 J. B. Martin and D. B. Chaffin, Biomechanical computerized simulation of human strength in sagittal plane lifting. AIIE Trans. 4 (1972), 19-28.

16 T. K. Ghosh and W. H. Boykin, Dynamics of the human kip-up maneuver. ASME J. Dynamic Systems, Measurem. Control 97 (1975), 196-201.

17 B. A. Murtagh and W. R. Sargent, A constrained minimization method with quadratic convergence, in R. Fletcher (ed.), Optimization, (London: Academic Press, 1969).

# Index